Metabolic Cardiomyopathy

Edited by
H. Böhles and A. C. Sewell, Frankfurt

With contributions by
M. Beck, H. Böhles, V. Hesse, C. Kampmann, E. Kauf, W. Kienast,
G. Mall, T. Marquardt, E. Mengel, H. Przyrembel, R. Santer,
A. A. Schmaltz, A. C. Sewell, W. Sperl, C. F. Wippermann

Second revised edition

medpharm *Scientific Publishers Stuttgart*

Editors:
Prof. Dr. H. Böhles
Dr. A. C. Sewell
Department of General Paediatrics
Johann-Wolfgang-Goethe-University
Theodor-Stern-Kai 7
D-60590 Frankfurt/Main
Germany

No responsibility is assumed by the publisher for any injury and/or damage to persons or property as a matter of products liability, negligence or otherwise, or ideas contained in the material herein. Because of rapid advances in the medical sciences, the publisher recommends that independent verification of diagnosis and drug dosages should be made.
The use of general descriptive names, trade names, trademarks, etc. in a publication, even if not specifically identified, does not imply that these names are not protected by the relevant laws and regulations.

Bibliographic information published by Die Deutsche Bibliothek
Die Deutsche Bibliothek lists this publication in the Deutsche Nationalbibliografie; detailed bibliographic data is available in the Internet at http://dnb.ddb.de.

ISBN 3-88763-104-8

All rights reserved. No part of this publication may be translated, reproduced, stored in a retrieval system, or transmitted, in any form or by any means, electronic, mechanical, photocopying, microfilming, recording or otherwise, without permission in writing from the publisher.

© 2004 medpharm Scientific Publishers
Birkenwaldstr. 44, D-70191 Stuttgart
Printed in Germany
Typesetting: TEBITRON GmbH, Gerlingen
Printing: Maisch & Queck, Gerlingen
Cover design: Atelier Schäfer, Esslingen

Foreword

About thirty years ago, in the 1972 edition of the renown "Nadas Textbook of Paediatric Cardiology", only about 16 of the 750 pages were dedicated to cardiomyopathies. Their cause could only be attributed unspecifically to some disease suggestions. In the work up of chronic cardiomyopathies it was recommended to take a muscle biopsy for the exclusion of "glycogen disease" and to fractionate plasma and urine amino acids to detect metabolic disease. Since these recommendations, tremendous advances in our understanding of metabolic diseases and their diagnostic approach have developed. Today, cardiomyopathies can to a large extent be attributed to inborn metabolic causes. So, we have learned for instance that patients with defects of very long-chain fatty acid oxidation almost always present with cardiomyopathy and lysosomal defects, like Fabry disease, up to recently only connected with the dermatological sign of angiokeratoma, have their own cardiac manifestation. This compilation of articles reviews the present understanding of cardiomyopathies of metabolic origin. However, we are well aware that new knowledge is accumulating rapidly and such a book may well not contain the latest facts at the moment of its appearance. It was therefore not our intention to give a comprehensive review of metabolic cardiomyopathies, but to provide practical information for physicians who have to take diagnostic and therapeutic decisions in patients with cardiomyopathy.

H. Böhles   Frankfurt/Main, September 2003

Table of Contents

Foreword V

1 Morphological presentation of cardiomyopathy 1
G. Mall

1.1 Introduction 1
1.2 Differential diagnosis of hypertrophic cardiomyopathy 1
1.3 Differential diagnosis of dilated cardiomyopathy 4

2 Possibilities and frontiers of myocardial biopsy in childhood 11
A. A. Schmaltz

2.1 Biopsy techniques 11
2.2 Histopathological diagnostics 12
2.3 Indication 13
2.4 Success rate and complications 14
2.5 Conclusions 16

3 Disturbances of the carnitine system as a cause of cardiomyopathy 17
H. Böhles

3.1 Developmental aspects of myocardial energy metabolism 17
3.2 The carnitine plasma membrane transport 17
3.3 Myocardial fatty acid oxidation 18
3.4 Carnitine plasma membrane transport defect (systemic carnitine deficiency) . 18
3.5 The carnitine palmitoyltransferase (CPT) system 19
3.5.1 CPT 1 19
3.5.2 CPT 1 deficiencies 19
3.5.3 CPT 2 deficiencies 20
3.5.4 Carnitine translocase deficiency 21
3.5.5 Secondary carnitine deficiency 22

4 Defects in long-chain fatty acid oxidation as a cause of cardiomyopathy 25
H. Przyrembel, A. C. Sewell

4.1	Long-chain acyl-CoA dehydrogenase (LCAD) deficiency 29	4.4	Trifunctional protein deficiency 31
4.2	3-Hydroxyl-CoA dehydrogenase (LCHAD) deficiency 29	4.5	Very-long-chain acyl-CoA dehydrogenase (VLCAD) deficiency 31
4.3	2, 4 Dienoyl reductase deficiency ... 30	4.6	Long-chain 3-ketothiolase deficiency 31

5 Cardiomyopathy in β-ketothiolase deficiency 35
V. Hesse, A. C. Sewell, H. Böhles, H. Haberland, B. Middleton, B. Fiedler, H. Förster, W. Jänisch

5.1	Case report 37	5.1.3	Neurological status 40
5.1.1	Family history 37	5.1.4	Cardiological status 40
5.1.2	Patients 37		

6 Cardiac involvement in glycogen storage diseases 47
R. Santer, K. Ullrich

6.1	Biochemical bases 47	6.2.1	Disorders with cytosolic glycogen storage 52
6.1.1	Glycogen storage diseases 47		
6.1.2	Disorders with vacuolar glycogen storage 51	6.2.2	Disorders with vacuolar glycogen storage 58
6.2	Clinical aspects 52		

7 Cardiomyopathies and mitochondrial defects of oxidative energy metabolism 67
W. Sperl

7.1	Characteristics of the oxidative phosphorylating system 68	7.5	Classification of mitochondrial cardiomyopathies 73
7.2	Aging presbycardia 69	7.5.1	Inheritance 73
7.3	Cardiomyopathies as a consequence of a secondary damage of the OXPHOS system (secondary mitochondrial cardiomyopathy) 70	7.5.2	Enzyme defects 75
		7.5.3	Characterization of the gene defect .. 75
		7.5.4	Mitochondrial syndromes with heart involvement 76
7.4	Primary mitochondrial cardiomyopathies 71	7.6	Disturbances of cardiac rhythm in defects of the OXPHOS system 76

7.7	Frequency of cardiac involvement in mitochondrial (encephalo-) myopathies 77	7.9	Therapy 79
7.8	Diagnosis of mitochondrial cardiomyopathies 77	7.10	Summary and outlook 79

8 Pericardial effusion refractive to therapy in an infant with the carbohydrate-deficient glycoprotein syndrome 85
W. Kienast, F. Walter, K. Heyne

9 Cardiomyopathy in congenital disorders of glycosylation (CDG) 87
J. Gehrmann, H. Böhles, T. Marquardt

9.1	Introduction 87	9.3	Discussion 94
9.2	Case reports 88	9.4	Conclusion 96

10 Cardiovascular changes in the mucopolysaccharidoses 99
C.-F. Wippermann, M. Beck, D. Schranz, R. Huth, I. Michel-Behnke, B.-K. Jüngst

10.1 Changes of the coronary arteries .. 100	10.6.1 Pathoanatomical results 105
10.1.1 Pathoanatomical results 100	10.6.2 Clinical findings 106
10.1.2 Clinical results 101	10.7 Therapeutical influence on the cardiovascular alterations 109
10.1.3 Discussion 101	
10.2 Alterations of other arterial vessels . 102	10.7.1 Bone marrow transplantation 109
10.3 Arterial hypertension 102	10.7.2 Cardiac valve replacement 109
10.4 Myocardial changes 103	10.8. Mortality as a consequence of cardiovascular causes 110
10.4.1 Pathoanatomical results 103	
10.4.2 Clinical observations 103	10.9 Summary 110
10.5 Changes in the conduction system .. 104	10.10 Conclusions 111
10.6 Alterations of the heart valves 105	

11 Cardiovascular involvement in Gaucher disease 113
E. Mengel

11.1 Introduction 113	11.2.3 Heart valve calcification 114
11.2 Heart 114	11.3 Cor pulmonale 115
11.2.1 Pericarditis 114	11.4 Conclusion 115
11.2.2 Myocardial infiltration 114	

12 Selenium deficiency in children with cystic fibrosis and phenylketonuria-metabolic and echocardiographic findings during sodium selenite therapy   117

E. Kauf, L. Vogt, J. Seidel, K. Winnefeld, H. Richter, H. Vogl, H. Dawczynski, A. Forberger, D. Schlenvoigt

12.1 Patients and methods .... 117 | 12.2 Results and discussion 118

13 Fabry disease – a progressive multisystemic lysosomal storage disorder   123

M. Beck

13.1 Pathophysiology 123
13.2 Symptoms 124
13.2.1 Pain 124
13.2.2 Skin 125
13.2.3 Autonomic dysfunction 125
13.2.4 Eyes 126
13.2.5 Heart 126
13.2.6 Kidney 126
13.2.7 Cerebrovascular system 126
13.3 Genetics 127
13.4 Treatment 127

14 The Anderson-Fabry disease associated cardiomypathy   133

C. Kampmann

14.1 Introduction 133
14.2 Cardiac manifestation 134
14.2.1 Structural changes of the myocardium 139
14.2.2 Changes in systolic function 139
14.2.3 Changes in diastolic function 140
14.2.4 Valvular involvement 142
14.2.5 Involvement of the conduction system 142
14.2.6 Coronary artery disease 143
14.3 "Cardiac variant" of AFD 144
14.4 Gender related differences 144
14.5 Onset and progression of the cardiomyopathy 145
14.6 Clinical features of cardiac involvement 146
14.7 Cardiac involvement in children with AFD 146
14.8 Natural history and death 146
14.9 Treatment 147
14.9.1 Effects of enzyme replacement therapy on the heart 147
14.9.2 Indication for treatment 150

15 Laboratory diagnosis of metabolic diseases presenting with cardiomyopathy 153
A. C. Sewell

15.1 Defects in mitochondrial
long-chain fatty acid oxidation 153
15.2 Defects in carnitine metabolism ... 154
15.3 Disorders of the respiratory chain .. 155

15.4 Disorders of complex
carbohydrate metabolism 155
15.5 Organoacidopathies 157
15.6 Others 157

Authors .... 163 | **Index** 165

1 Morphological presentation of cardiomyopathy

G. Mall

1.1 Introduction

According to their morphology, cardiomyopathies are classified into three forms: dilated, hypertrophic and restrictive [14,15]. Although a WHO board of experts suggested that the concept of cardiomyopathy should cover only the idiopathic (primary) form [14], this restrictive definition has proved to be unsatisfactory for clinical cardiology and has not found general acceptance. Heart diseases of well-known cause e.g. alcohol-induced are still commonly referred to as cardiomyopathies [9, 15]. The purpose of the present paper is to discuss the differential diagnosis of dilated as well as hypertrophic cardiomyopathies. The restrictive type, occurring among adults in Central Europe almost exclusively as a form of Löffler myocarditis, will not be dealt with [14].

1.2 Differential diagnosis of hypertrophic cardiomyopathy

Hypertrophic cardiomyopathies are marked by concentric biventricular hypertrophy. Transitions to a dilated form have been described, however, dilatation appears subsequent to hypoxia in most cases, with co-existing intramyocardial microarteriopathy. Macromorphologically, more than 90% of all cases show asymmetric hypertrophy, usually including the septum ventriculorum (asymmetric septum hypertrophy). If the degree of asymmetric hypertrophy is highest in relation to the left ventricular outflow tract an obstructive form of cardiomyopathy may occur. Apical types are seldom found. At autopsy great variation in macroscopic and microscopic features is revealed. Heart weights range between 400 g and more than 1000 g. According to our own observations, the lowest measured was 380 g and belonged to a young male patient aged 18, who died rather unexpectedly probably from rhythmic dysfunction. Besides disturbance in the rhythm, more severe diastolic dysfunctions (compliance disturbance) may appear even in the presence of a low rated hypertrophy. One personal observation refers to a young female patient, 21 years of age, whose heart weighed no more than 450 g at autopsy and who also showed no sign of asymmetric hypertrophy. Histologically, the typical alterations of hypertrophic cardiomyopathy (with positive family history) were found.

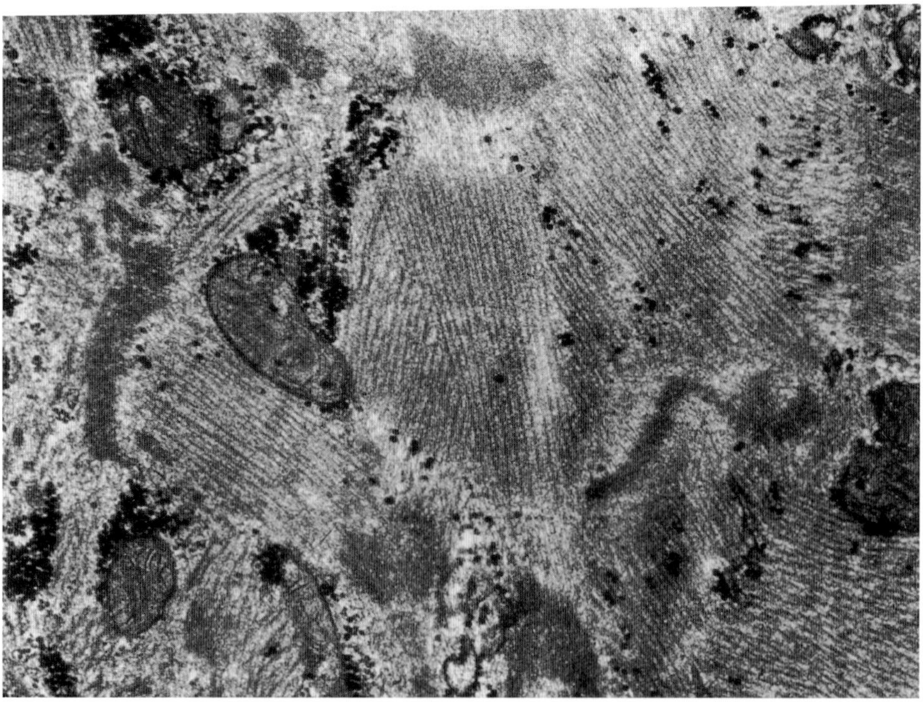

2 Differential diagnosis of hypertrophic cardiomyopathy

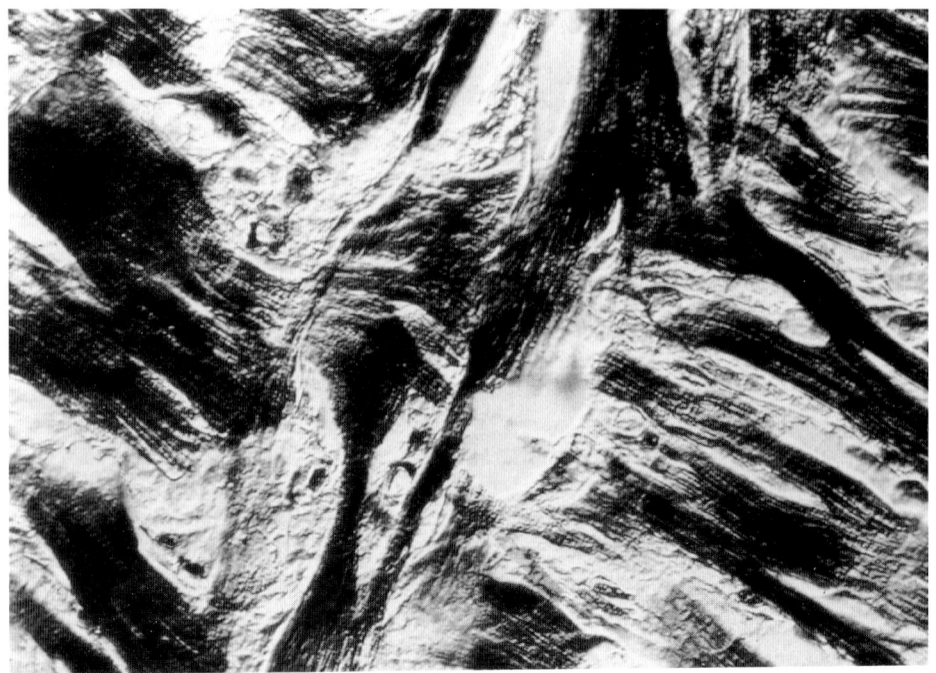

In addition to asymmetric hypertrophy, some other irregularities in the architecture of the ventricles are most common. The rough texture of the inner ventricular surface often appears to be atypical, the trabecular network has another differentiation, and the papillar muscles show anomalous forms. Under the microscope, many ventricular sections present myocyte texture disturbance with whirling, and some muscle fibres appear to have vertical orientation to one another (Fig. 1). This finding shows variations in each heart and differs with each patient, yet it bears no specific value, since slight myocyte whirls occur even in the normal heart. When hypertrophic cardiomyopathy is suspected, it is necessary to prepare several histological sections and quantify the texture disturbance [13]. Evidence of relevant texture deviation is most likely to be given in holoptic transverse sections of the septum ventriculorum. In diagnoses using very small myocardial biopsies, the texture criteria in hypertrophic cardiomyopathy - compared to autopsy - has no specific relevance.

Differential diagnosis of hypertrophic cardiomyopathy refers to heart diseases causing hypertrophy or wall enlargement in the ventricle without producing a higher graded dilatation. Hypertrophic cardiomyopathy associated with neurological or myopathic affections, as in Friedreich ataxia and myotonic dystrophy [4] are unproblematic from a differential diagnostic viewpoint. Occasionally, in diagnosing cardiac amyloidosis some problems will have to be faced (Fig. 2), and in order to clarify the situation when other diagnostic measures (e.g. rectal biopsy) fail, endomyocardial biopsy may be requested [16]. Thesaurismosis can also appear as hypertrophic cardiomyopathy. The X-chromosomal recessively transmitted Fabry disease (Fig. 3), a sphingolipidosis characterized by ceramide-trihexoside deposits (due to α-galactosidase deficiency) in heart muscle cells, renal tubulus epithelium, mesenchymal cells and in the endothelium may present as a cardiac form, which does not imply kidney participation, presence of angiokeratoma or retinopathy [3, 12]. Most glycogenoses with cardiac involvement become manifest in infancy, the shape of the heart and ventricular wall enlargement showing similarities to hypertrophic cardiomyopathy. In mucopolysaccharidoses, enlarged ventricles are also observed. When diagnosing therausismosis at autopsy, it should be taken into account that the accumulated substances can no longer be determined since they have already been dissolved by the conventional fixation and embedding methods. With an adequate technique, the diagnostic sensitivity using endomyocardial biopsies to determine amyloidosis or thesaurismosis is high because the changes are homogenously distributed in the myocardium [2].

Finally it must be mentioned that cases of hypertrophic cardiomyopathy can also be seen in generalized mitochondriopathies. Under the light microscope, this disease is generally strik-

Fig.1: Light microscopic (above) and electron microscopic (below) representation of texture disturbance in hypertrophic cardiomyopathy. The light microscope reveals branching, partially vertically directed myocytes, the electron microscope shows in turn irregular orientation of myofilaments and Z-shaped stripes. In between are mitochondria.

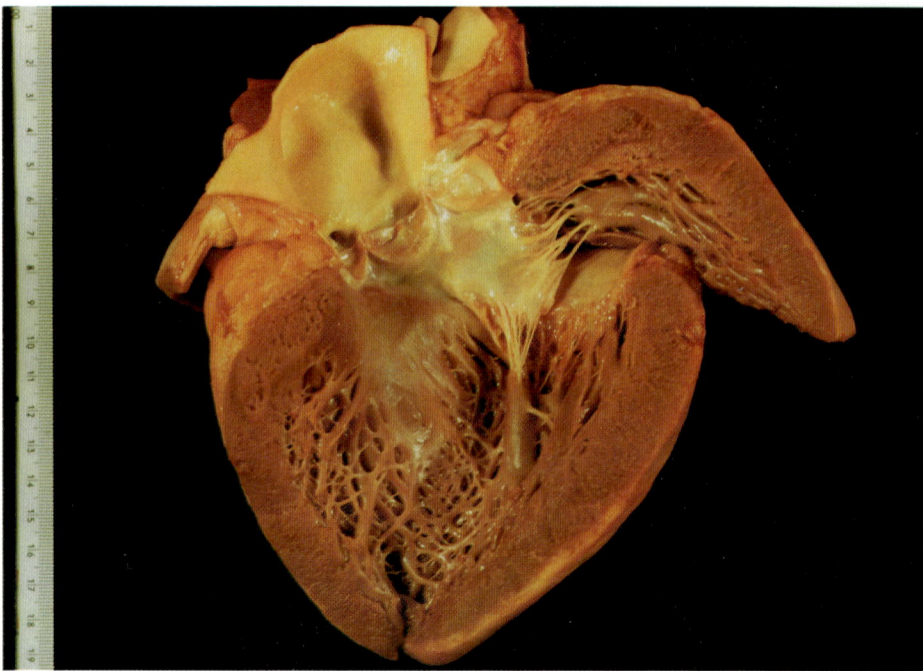

Fig. 2: Cardiac amyloidosis showing a bulge in septum ventriculorum and an enlarged ventricular wall misleading clinically to hypertrophic cardiomyopathy

ing because of enlarged mitochondria dispersing myofibrils. With the electron microscope atypical mitochondria with disturbance in the architecture of the inner mitochondrial membranes (cristae mitochondriales) are revealed [18]. Even metabolic disturbances such as carnitine deficiency may occur as hypertrophic cardiomyopathy, but specific histological or ultrastructural modifications have not been described. In a number of cases a fine-droplet intermyofibrillar fatty degeneration has been observed [4].

1.3 Differential diagnosis of dilated cardiomyopathy

The main morphological feature of dilated cardiomyopathy is ventricular and atrial dilatation with compensatory myocyte hypertrophy (due to an increase in wall tension). In spite of this myocyte hypertrophy, the ventricular walls are only slightly widened or even narrowed. Endocardial fibrosis gene-

Fig. 3: Light microscopic representation of a case of Fabry disease (semithin section, embedding in Araldite). The muscle cells show accumulation of lamellar and myelin figures representing ceramide trihexoside.

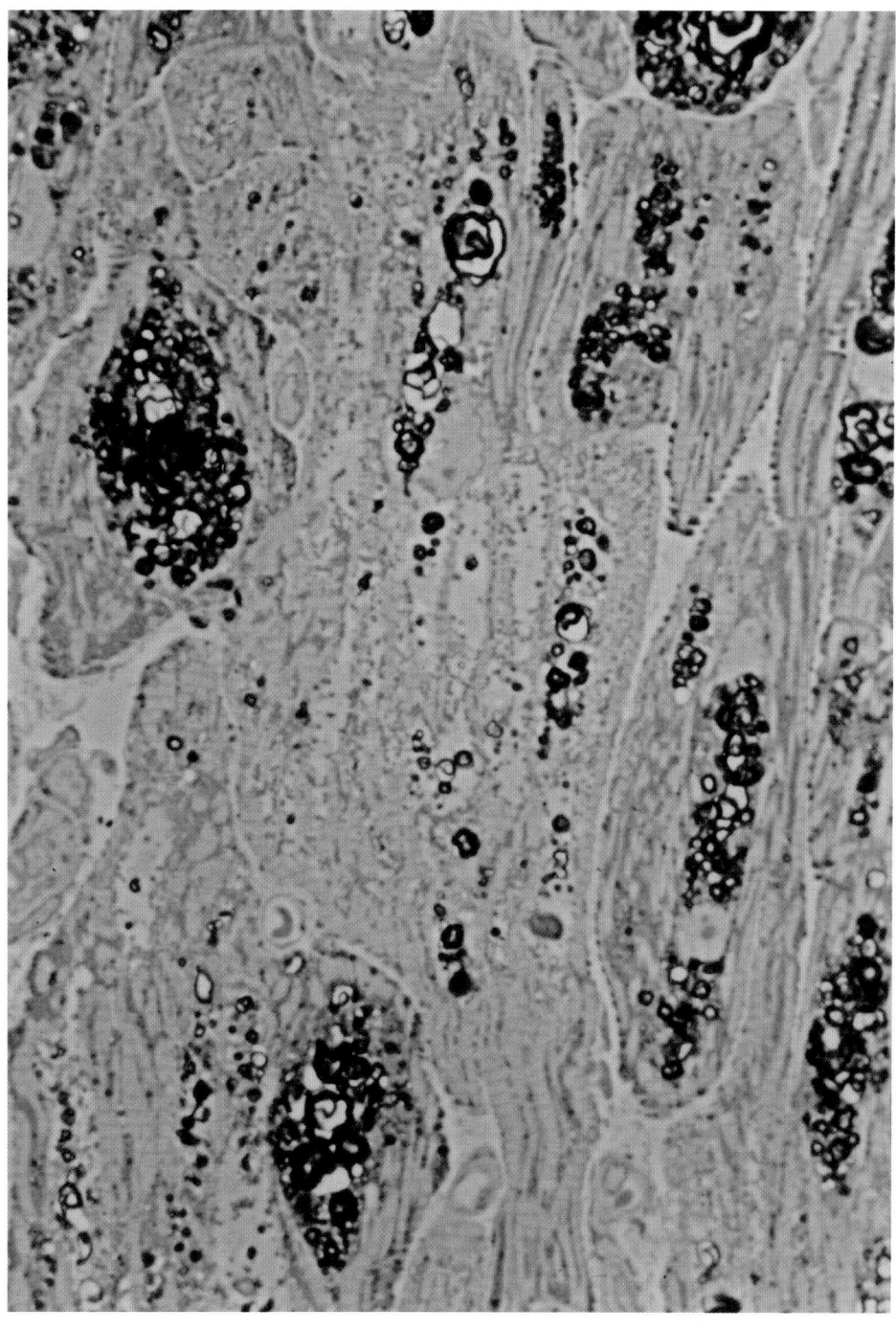

rally occurs within dilated ventricles, and in infancy, transition to restrictive cardiomyopathy of the fibroelastosis type will result. Endocardial fibrosis is interpreted as a reaction to the abnormal drawing tracts.

Dilated cardiomyopathies can be "identified" (e.g. dilated cardiomyopathy within the strict WHO definition) as familial or toxic (alcohol, anthracycline), and can also appear in collagenoses, chronic infections and vasculites. Neurological and myocardial diseases occur not only as hypertrophic myopathies (see above) but also in their dilated forms (as in Friedreich ataxia, Duchenne muscular dystrophy, dystrophic myotonia and mitochondrial myopathies).

Morphologically, the dilated hearts present considerable myocyte hypertrophy, frequently showing marked interstitial myocardial fibrosis and sometimes even a reparative fibrosis (scars). The ultrastructure is characterized by myofibril reduction and mitochondria of different form and size (not to be confused with mitochondrial cardiomyopathy). Ventricular function correlates with the myofibrillar content. Structural myocardial alterations are unspecific correlates of cardiac insufficiency, and hence might not show histological evidence of dilated cardiomyopathy [5, 10, 11].

Alcohol-induced cardiomyopathies cannot be distinguished from idiopathic cardiomyopathies. On the other hand, anthracycline-induced cardiomyopathies are often characterized by a marked dilatation of the sarcoplasmic reticulum

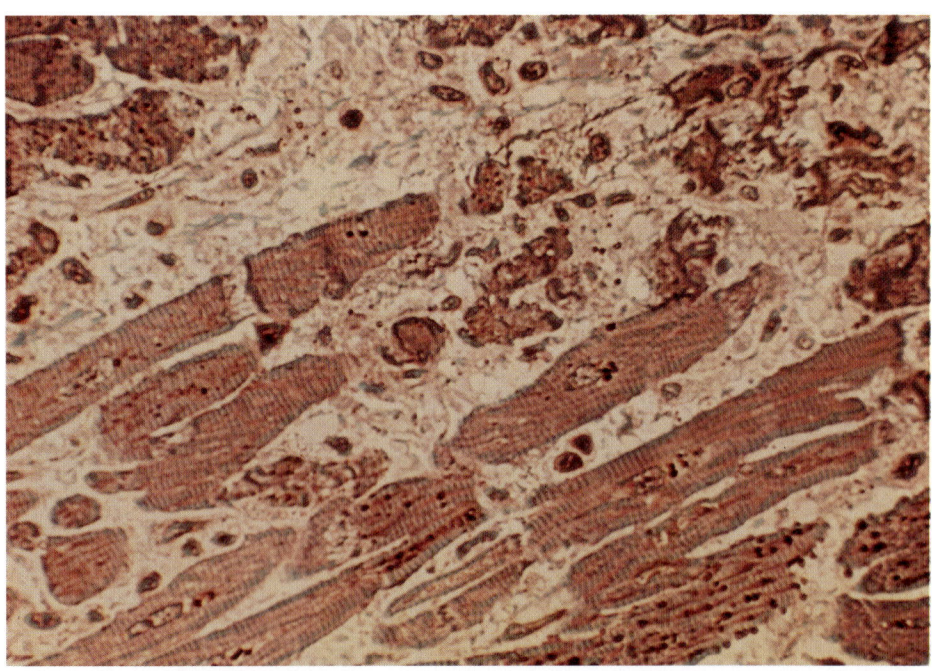

Fig. 4: Myoctolysis in virus myocarditis. The upper part of the image shows well-preserved myocytes with distinctively discernable transverse striations. The lower part shows degenerating myocytes with detaching cell membranes and myofibrillar hypercontractions.

and by chromatin clumping in the muscle cell nuclei [2].

According to personal investigations, dilated cardiomyopathies produced by a neuromuscular affection (as in Friedreich ataxia, myotonic dystrophy, Duchenne muscular dystrophy) reveal a certain particularity: by the final stage of cardiac insufficiency about 80% of the ventricular myocytes will be destroyed. In contrast, the majority of idiopathic dilated cardiomyopathies show an increase in the number of myocytes due to hyperplasia (through which loss of myocytes is compensated). Mitochondrial cardiomyopathies also occur as dilated forms.

The differential diagnosis between infectious and noninfectious dilated cardiomyopathies is of great clinical relevance and actual scientific interest. Enteroviruses are the most frequent cause of an infectious cardiomyopathy. Relevant causative agents for myocarditis are Coxsackie viruses of groups A and B, and above all Coxsackie virus B. Histologically, virus-induced myocardites are characterized by lymphatic infiltration, oedema, myocytolysis and later by reparative processes (Fig. 4). Lymphocytic infiltration represents an immunological response to virus infected myocytes. The first wave of infiltrating lymphocytes consists of natural killer cells, the second of cytotoxic T-lymphocytes and T-helper cells [6, 9, 10].

According to experimental findings in the mouse, the acute stage of virus-induced myocarditis ends already 14 days after onset. Although in humans protracted forms of progression are found (ongoing myocarditis), if myocarditis is clinically suspected, a biopsy should be obtained as soon as possible. According to the Dallas criteria [1], in diagnosing histologically a myocarditis, evidence of myocytolysis and inflammatory infiltration (Fig. 5) must exist. Lymphocytic infiltrations with no muscle fibre necrosis are classified as borderline myocarditis (Fig. 6). The sensitivity of endomyocardial biopsies for evidence of virus-induced myocarditis depends on the extent of inflammatory lesions. If only a small part of the myocardium is involved (e.g. < 10%), there is virtually no chance of finding a myocarditis in a biopsy. However, the importance of clinically less developed myocarditis might not be too great. In extensive (clinically relevant) lesions which include 30 to 40% of the myocardium, security in diagnosing is likely to be acceptable, if at least five endomyocardial biopsies are obtained.

According to clinical observations, a dilated cardiomyopathy will arise in part in some patients who have previously had a virus-induced myocarditis [7]. Despite the temporal relationship, endomyocardial biopsies in cases of chronic dilated cardiomyopathy seldom show an active myocarditis (corresponding to the Dallas criteria). Recent findings in molecular biology and immunochemistry have furnished some new viewpoints. In about 20% of patients with cardiomyopathy, enteroviral RNA by polymerase chain reaction or in situ hybridisation has been determined [7]. Owing to sequential homologies of different enteroviruses, it is possible to determine enteroviral RNA-specific groups, so that evidence in molecular biology will include all the Coxsackie viruses of groups A and B. Persistency in viral infection can be explained by masked infected cells opposing the immune system.

A change in the viral replication of infected cells, for example, could cause such an alteration in cell metabolism, that the cell would not code MHC class

I molecules by which the cytotoxic T-lymphocytes would no longer be able to identify the infected cells. Furthermore, defective mutants might have lost coding for viral coat proteins by which the immune system no longer recognizes the infected cells [9].

Experimental findings and single observations on human endomyocardial biopsies draw suspicion that enteroviral heart infections are associated in many cases with infiltration of T-lymphocytes; but the number of T-lymphocytes is hereby much more reduced than in cases of acute virus myocarditis [8]. On the other hand, the numeric relation of T-lymphocytes and infected cells in the persisting stage is markedly raised in favour of T-lymphocytes (therefore, asserting the fact that virus infected cells in their phase of persistency are not effectively eliminated by the immune system) [8]. For the time being, no definition for the affections known as chronic myocarditis has been found by common international agreement. In our immunochemical experiments on the murine model of myocarditis and present experience in endomyocardial biopsy in humans, the critical limit for the number of infiltrating T-lymphocytes in the myocardium might be about 10 to 30 per square millimetre.

If a diagnostic endomyocardial biopsy is requested to determine chronic dilated cardiomyopathy, in addition to hi-

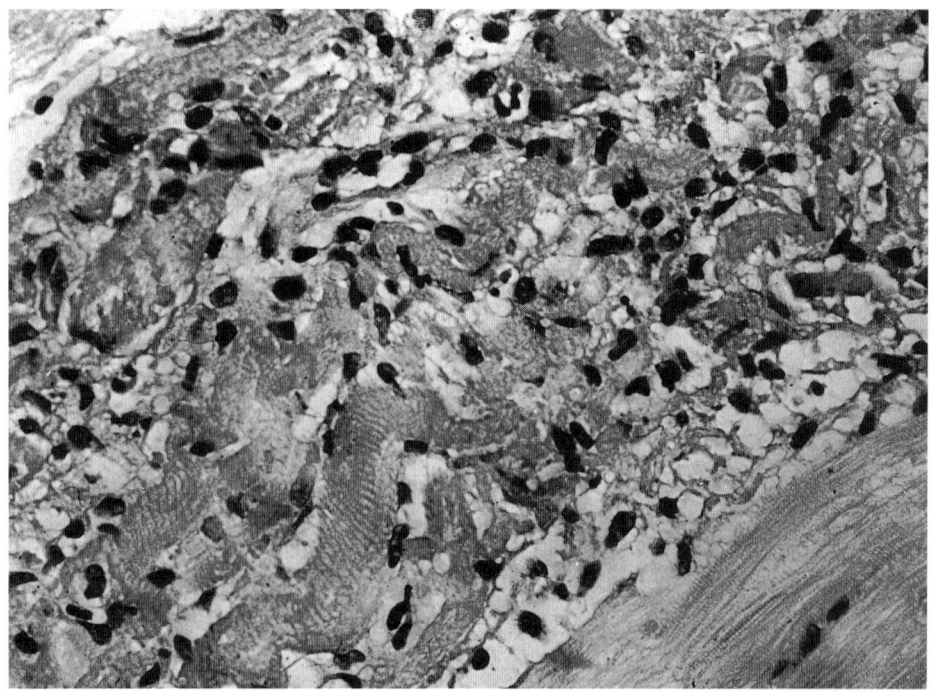

Fig. 5: An acute virus myocarditis in a left ventricular endomyocardial biopsy. On the lower right side preserved myocytes are seen. The remainder shows necrobiotic myocytes and lymphocytic infiltration of the surrounding area.

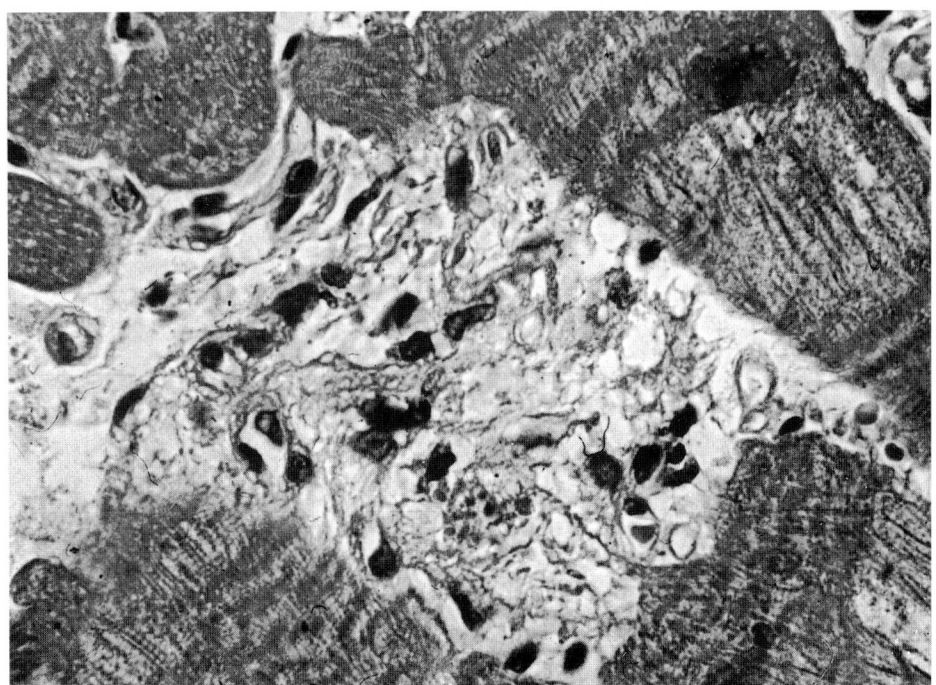

Fig. 6: A left ventricular endomyocardial biopsy of borderline myocarditis with well preserved myocytes, no myocytolyses and focal lymphocytic infiltration (centre).

stological analysis further investigation in molecular biology and immunochemistry should follow [17].

In conclusion, it must be emphasized that some of the modern methods can be applied merely to native tissue (e.g. frozen sections). Molecular biological analyses may be obtained from formalin fixed samples, but their sensitivity is somewhat limited. Immunochemistry meanwhile uses antibodies that can recognize antigens on paraffin section. Hence, native samples are no longer necessarily requested.

Acknowledgements

The excellent photographic work and technical assistance of Harald Derks, Dieter Schneider, Rita Haack and Edit Jung is gratefully acknowledged.

References

[1] Aretz, H. T.: Myocarditis: The Dallas criteria. Hum. Path. **18**, 619-624 (1987).
[2] Billingham, M. E.: Role of endomyocardial biopsy in diagnosis and treatment of heart disease, in: M. D. Silver (ed.): Cardiovascular Pathology, p. 1465, Churchill Livingstone (1991).
[3] Erdmann, E., H. D. Bolte, B. E. Strauer, G. Hübner: Myokardiale Beteiligung bei Morbus Fabry. Dtsch. med. Wschr. **105**, 168 (1980).
[4] Ferrans, V. J.: Metabolic and familial diseases, in: M. D. Silver (ed.): Cardiovascular Pathology, p. 1073, Churchill Livingstone (1991).
[5] Frenzel, H., M. Kasper, H. Kuhn, G. Lösse, G. Reifschneider, W. Hort: Licht- und elektronenmikroskopische Befunde in Früh- und Spätstadien der Herzinsuffizienz. Untersuchungen an Endomyokardbiopsien von Patienten mit latenter (LCM) und dilatativer (DCM) Kardiomyopathie. Z. Kardiol. **74**, 135-143 (1985).
[6] Kandolf, R., P. H. Hofschneider: Molecular cloning of the genome of a cardiotropic coxsackie B 3 virus: Full-length reverse-transcribed recombinant cDNA generates infectious virus in mammalian cells. Proc. nat. Acad. Sci. (Wash.) **82**, 4818-4822 (1985).
[7] Kandolf, R., P. H. Hofschneider: Viral heart disease. Semin. Immunopath. **11**, 1-13 (1989).

[8] Klingel, K., C. Hohenadl, A. Canu, M. Seemann, G. Mall, R. Kandolf: Ongoing enterovirus-induced myocarditis is associated with persistent heart muscle infection: Quantitative analysis of virus replication, tissue damage and inflammation. Proc. nat. Acad. Sci. **89**, 314-318 (1992).

[9] Kuhn, H., F. Loogen: Die Wirkung von Alkohol auf das Herz einschließlich der Alkoholkardiomyopathie. Internist **19**, 97 (1978).

[10] Kunkel, B., M. Schneider, W. D. Kober, R. Hopf, M. Kaltenbach: Die Morphologie der Myokardbiopsie und ihre klinische Bedeutung. Z. Kardiol. **71**, 787 (1982).

[11] Mall, G., F. Schwarz, H. Derks: Clinicopathologic correlations in congestive cardiomyopathy. Virchows Arch. A **397**, 67 (1982).

[12] Mall, G., F. Schwarz, H Zebe, R. Waldherr: Cardiale Form des Morbus Fabry ohne Hautbeteiligung. Verh. Dtsch. Ges. Path. **66**, 484 (1982).

[13] Maron, B. J., W. C. Roberts: Quantitative analysis of cardiac muscle cell disorganization in the ventricular septum of patients with hypertrophic cardiomyopathy. Circulation **59**, 689 (1979).

[14] Report of the WHO/ISFC task force in the definition and classification of cardiomyopathies. Brit. Heart. J. **44**, 672-673 (1980).

[15] Roberts, W. C., V. J. Ferrans: Pathologic anatomy of the cardiomyopathies. Hum. Pathol. **6**, 287 (1975).

[16] Sedlis, S. E., J. E. Saffitz, V. S. Schwob, A. S. Jaffee: Cardiac amyloidosis simulating hypertrophic cardiomyopathy. Am. J. Cardiol. **53**, 969 (1984).

[17] Schultheiss, H. P.: Immunsuppressive Therapie bei Myokarditis und dilatativer Kardiomyopathie? Internist (Berl.) **33**, 650-662 (1992).

[18] Schwartzkopff, B., H. Frenzel, G. Breithardt, M. Dekkert, Lösse, K. Toyka, M. Borgrefe, W. Hort: Ultrastructural findings in endomyocardial biopsy of patients with Kearns Sayre syndrome. J. Amer. Coll. Cardiol. **12**, 1555-1558 (1988).

2 Possibilities and frontiers of myocardial biopsy in childhood

A. A. Schmaltz

Transvascular endomyocardial biopsy (EMB) is a well documented diagnostic method in adults. In 1982 the Stanford group [2] reported on over 4000 patients who had undergone myocardial biopsy for suspected rejection after heart transplantation, for antrazycline-induced cardiomyopathy and for myocarditis.

In the beginning the diameter of bioptomes (1.8-2.5 mm), which had to be inserted via 9 F pilot catheters, prevented the use of this technique in childhood. The new generation of bioptomes is much smaller so that they can be inserted by means of a 6 F or 7 F pilot catheter. Based on our own experience with this technique in 100 infants and children, the possibilities and frontiers of myocardial biopsies are discussed.

2.1 Biopsy techniques

After the time when needle and punch biopsies were used, the described EMB is performed at the end of routine heart catheterization for the evaluation of haemodynamics, left ventricular function or coronary blood supply. In this procedure a thin catheter, using a guide wire if necessary, is introduced into the punctured inguinal vessel and extended to the right ventricle or via the oval foramen to the left ventricle of the heart. When the bioptome reaches the top of the catheter, the jaws are opened and gently pressed against the endocard. The jaws are then closed and the bioptome withdrawn. It is almost impossible to influence the point of biopsy. In the right ventricle it is important to position the bioptome at the tip of the heart and not in the outflow tract. In the left ventricle the area between the papillary muscles is normally reached. General anaesthesia is not necessary and only a small amount of time and material is needed.

The biopsy technique was recently modified and performed under ultrasound guidance [7]. This allows the localisation of the bioptome at the moment of myocardial contact and immediately allows the diagnosis of complications like myocardial perforation or the development of pericardial haemorrhage. Another modification concerns the use of a Mullins-sheath which is introduced via a balloon catheter, positioned at the tip of the right ventricle [4]. Needless to say it is more difficult to position the bioptome in the heart of very small infants, because of the limited space necessary to alter direction and the instrument can easily retreat back to the right atrium. Figure 1 shows the age and weight distribution of our patients from 1 month up to 22 years and 3 to 67 kg.

2.2 Histopathological diagnostics

Because the amount of bioptically obtained tissue is very small and only minor pathological alterations of the myocard may be present, a maximum of obtainable investigational procedures have to be used by the pathologist. We normally perform 3 to 5 biopsies, which are prepared for light- and electron microscopic analysis. Another tissue sample is maintained at -70 °C for immuno-histological investigations. The presence of antimyolemmal and antisarcolemmal antibodies, adherent to myocytes is investigated by direct or indirect immunofluorescence [6]. Glycogen and iron deposits can be demonstrated histochemically. The technique of in situ genetic hybridisation, i.e. the recombination of complementary nucleic acid sequences allows the demonstration of a virus genome in the myofibrils [3]. This fact can be used as evidence that the myocardial cell has been in contact with a specific virus, for instance with CMV, HSV, EBV, Coxsackie virus and others. However all these results have to be interpreted in context with the results from other classical methods of investigation such as light and electron microscopy, biochemical analysis and the entire clinical picture.

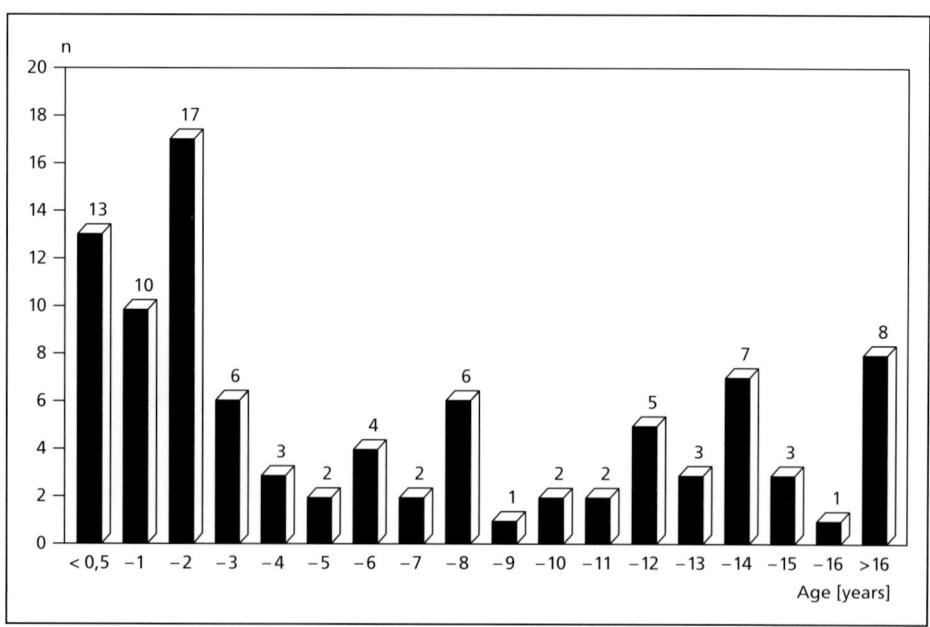

Fig. 1: Age of patients undergoing endomyocardial biopsy. n = 95; 52 male; 43 female; average age = 6.08 ± 6 years.

2.3 Indication

1. The largest group of patients with EMB was 52 patients with a dilated left ventricle and a myocardial function.

Endocardial fibroelastosis is a disease of infancy. The left ventricle is lined with a porcelain-like endocardial thickening, representing an inner protective shell. In these cases the biopsy is extremely difficult to perform because the jaws slip from the tissue; however histology showing endocardial thickening is absolutely specific in these cases.

The different forms of myocarditis are a difficult diagnosis for the pathologist. A test situation in the USA, when 16 different histological samples were sent to seven leading cardiopathologists, led to a surprising result: the frequency of the diagnosis myocarditis varied between 0 and 50% [10]. We therefore submit our histological samples to immunohistological investigation or to in vitro hybridisation of the viral genome, which in many cases gives a positive result in cases of chronic lymphocytic myocarditis.

In cases of dilated cardiomyopathy histopathology is uncharacteristic. In most cases hypertrophy of myofibrils, minor interstitial fibrosis and irregular cellular nuclei are found. Electron microscopy shows mitochondriosis and degenerative mitochondrial changes. Unfortunately, biopsy is not aetiologically conclusive for dilated cardiomyopathy. It is characteristic for the pathological changes of the heart that different pathomechanisms lead to identical histopathological changes. It cannot be concluded from the histology whether the alterations result from carnitine or selenium deficiency, athyroidism, fibromuscular dysplasia or Kawasaki syndrome. Only in the case of adriablastin induced cardiomyopathy and in haemosiderosis are histochemical analyses diagnostically helpful.

2. The second group of our biopsied patients consisted of 25 children with an inadequate muscular hypertrophy. These cases with typical sonographical structures of an obstructive cardiomyopathy are no longer an indication for EMB. The abnormal myofibrillar branching seen on electron microscopy is characteristic for hypertrophic cardiac myopathy. Biopsy was diagnostically helpful in one case in differentiating between hypertrophic cardiomyopathy and a tumour. The histology led us to an unsuccessful therapeutic trial. At autopsy our diagnosis made by biopsy could be proven. Cardiac tumours are virtually inaccessible to biopsy because they are not reached by the pincers or only surrounding healthy fibres are obtained. If tissue is obtained, it is diagnostic in all cases.

In contrast to this disease entity the concentric form of left ventricular hypertrophy represents a clear indication for EMB. The electron microscopy demonstration of lamellar phospholipid containing structures is diagnostic for Fabry disease. This is an X-linked inherited ceramide storage disease leading to a concentric form of hypertrophic cardiomyopathy as well as to general muscular hypotonia. These secondary forms of hypertrophic cardiomyopathy are much more important in infancy and childhood than in adults.

3. The last group comprised 8 patients with a different indication for EMB. Here we emphasise adriamycin induced cardiomyopathy. The extent of myofi-

brillar decolouration by means of the Berlin-blue reaction provides histological quantification of the damage and according to Bristow [1] represents the only method of prognostic value.

In Table 1 the different indications of EMB in childhood are given according to our own experience. At the same time this Table gives a certain evaluation of the method indicating it as "diagnostic" when a clear histological diagnosis was possible, as "helpful" when the histology was unspecific, but certain diagnoses could be excluded and "worthless" when histology could not contribute anything to the diagnoses. According to our results 11% of the biopsies were diagnostic, 75% were helpful and only 14% had to be considered as worthless.

2.4 Success rate and complications

The success rate of our biopsy series was 96%. Only three times was thrombotic material without myocardium obtained. In one infant the procedure had to be interrupted, when the leading catheter

Tab. 1: Indication and evaluation of endomyocardial biopsy in childhood.

Clinical symptom	Diagnosis	No. patients (died)	Value diagnostic	helpful	not helpful
▪ Dilated left ventricle	EFE	7 (3)	4	3	
	DCM post myocarditis	20 (8)		19	1
	DCM (resolving)	19 (2)	1	15	
	Myocarditis	9 (4)	3	6	
▪ Normal ventricle	RCM	4 (3)		3	1
▪ Excessive muscular hypertrophy	HCM	12 (3)		9	3
	Secondary hypertrophy	6 (1)		6	
	Storage disease	5 (1)	3	2	
	Tumour	2 (1)			2
▪ Miscellaneous	Hypoxic damage	4 (1)	4		
	Cytostatic damage	3	3		
	EFE in aortic stenosis	1	1		
	Bradycardia	1			1
▪ Not indicated		5			5
Total		95 (28)	11 (11%)	71 (75%)	13 (14%)

EFE = endocardial fibroelastosis; DCM = dilated cardiomyopathy; RCM = restrictive cardiomyopathy; HCM = hypertrophic cardiomyopathy kinked twice when the bioptome was protruded. Our success rate is well in accordance with that given in the literature [5]. The complication rate is of utmost importance for a diagnostic procedure. In a worldwide statistical survey of 2337 EMBs Richardson found severe complications in 1.67% [8].

Tab. 2: Diagnostic scheme of specific heart muscle diseases.

1. Anamnestic		
▪ Fetopathia diabetica ▪ ACTH	Adriblastin Haemosiderosis Cystic fibrosis Vit. B, E deficiency, selenium deficiency kwashiokor Fetal alcohol syndrome Kawasaki syndrome?	
2. Clinical Investigation	**Tests**	**Diagnosis**
▪ Skin	Chromosome analysis Skin biopsy	Noonan syndrome Lentiginosis
▪ Eyes		Leigh syndrome Kearns-Sayre syndrome Ceroid lipofuscinosis GM-gangliosidosis
▪ Blood pressure	Catecholamines Sonography Angiography	Phaeochromocytoma Fibromuscular dysplasia
▪ Neuromuscular symptoms	EMG/conduction velocity Carnitine	Duchenne muscular dystrophy Friedreich ataxia Carnitine deficiency
3. Laboratory Investigations	**Tests**	**Diagnosis**
	TSH/TBF Acidosis, Lactate Hypoglycaemia Vacuolated lymphocytes	Hypothyroidism Leigh syndrome GM-gangliosidosis Ceroid lipofuscinosis
4. Special Enzyme Assays	**Tests**	**Diagnosis**
	Acid 1,4-glucosidase Alpha-galactosidase A Respiratory chain enzymes	Pompe disease Fabry disease Leigh syndrome
5. Histology	**Tests**	**Diagnosis**
	Paracrystalline inclusions Glycogen Abnormal mitochondria "ragged red" fibres	Fabry disease Pompe disease Leigh syndrome

These complications were perforation with ensuing haemopericard formation, pericarditis, dysrhythmias, cerebral emboli, pneumothorax and sepsis. We experienced in our series of about 100 biopsies one pericardial haemorrhage following myocardial perforation. Yoshizato et al. [11] in contrast reported as serious complications three right ventricular perforations and one pneumothorax in their 53 biopsied patients.

2.5 Conclusions

Transvascular EMB is an established diagnostic method for infants and children with a low complication rate. In a series of diseases the specific pathohistological features allow a clear diagnosis. However, when the multitude of possible specific heart muscle diseases is taken into consideration, EMB represents the last in a sequence of possible diagnostic procedures. History, clinical investigation and biochemical investigation always have to precede the histological analysis. Table 2 may give some indications for the diagnostic approach [9].

References

[1] Bristow, M. R., J. W. Mason, M. E. Billingham, J. R. Daniels: Dose-effect and structure-function relationships in doxorubicin cardiomyopathy. Amer. Heart J., 102, 709 (1981).

[2] Fowles, R. E., J. W. Mason: Endomyocardial biopsy. Ann. Intern. Med. **97**, 885 (1982).

[3] Kandolf R., P. H. Hofschneider: Molecular cloning of the genome of a cardiotropic coxsackie B 3 virus: full-length reverse-transcribed recombinant eDNA generates infectious virus in mammalian cells. Proc. Natl. Acad. Sci. USA **82**, 4818 (1985).

[4] Leatherbury, L., R. S. Chandra, S. R. Shapiro, L. W. Perry: Value of endomyocardial biopsy in infants, children and adolescents with dilated or hypertrophic cardiomyopathy and myocarditis. J. Amer. Coll. Cardiol. **12**, 1547 (1988).

[5] MacKay, E. H., D. Pickering, W. A. Littler: Cardiac biopsy in childhood. Arch. Dis. Childh. **52**, 785 (1977).

[6] Maisch, B., W. Romen, P. Eigel, A. A. Schmaltz, V. Regitz, P. Deeg, G. Liebau, K. Kochsiek: Immunhistologische Befunde bei Perimyokarditis und dilatativer Kardiomyopathie. Verh. dtsch. Gsch. inn. Med. **90**, 1424 (1984).

[7] Mortensen, S. A., H. Egeblad: Endomyocardial biopsy guided by cross-sectional echocardiography. Br. Heart J. **50**, 246 (1983).

[8] Richardson, P. J.: Endomyocardial biopsy technique. In: H. D. Bolte (ed.): Myocardial biopsy. diagnostic significance. Springer, Berlin (1980).

[9] Schmaltz, A. A.: Diagnostik der Kardiomyopathien im Kindesalter. Zbl. Kind. **140**, 1 (1990).

[10] Shanes, J. G., J. Ghali, M. E. Billingham, V. J. Ferrans, J. J. Fenoglio, W. D. Edwards, C. C. Tsai, J. E. Saffitz, J. Issner, S. Furner, R. Subramanian: Interobserver variability in the pathologic interpretation of endomyocardial biopsy results. Circulation **75**, 401 (1987).

[11] Yoshizato, T., W. D. Edwards, E. T. Alboliras, D. J. Hagler, D. J. Driscoll: Safety and utility of endomyocardial biopsy in infants, children and adolescents: a review of 66 procedures in 53 patients. J. Amer. Coll. Cardiol. **15**, 436 (1990).

3 Disturbances of the carnitine system as a cause of cardiomyopathy

H. Böhles

3.1 Developmental aspects of myocardial energy metabolism

During cardiac development, alterations in energy metabolism depend upon oxygen delivery on the one hand and substrate availability on the other. During fetal life there is a strong reliance on aerobic glycolysis using essentially glucose and lactate as major energy sources. The energy metabolism of the mature heart is, in contrast, almost exclusively aerobic, using free fatty acids as main energy substrate. During hypoxia, energy metabolism is almost totally oriented towards anaerobic glucose utilisation, which, from a teleological point of view, has an oxygen sparing effect [1].

Basic knowledge concerning the energy substrates of the heart resulted from the pioneering work of Bing et al., who put the basis for the determination of substrate concentrations in arterial and venous cardiac blood by selective catheterization of the sinus venosus [2, 3]. They showed that free fatty acids, glucose and lactate are the main substrates of cardiac energy metabolism. At the same time it was shown that pyruvate, acetate, ketone bodies and amino acids can also be used by the heart as a source of energy. However, since these substances are present in only small concentrations, they are of no real importance, even when exogenously supplied [4]. It has been known for quite some time that free fatty acids are the essential energetic substrate for the heart based on measurements of a cardiac respiratory quotient (RQ) being < 1 [5].

3.2 The carnitine plasma membrane transport

The intracellular carnitine concentration is maintained by a plasma membrane carnitine transporter system. There are two distinct plasma membrane carnitine uptake mechanisms. In most tissues a high affinity transporter (OCTN2) with a Km ranging from 20 – 200 mM can be identified [6]. In the liver, by contrast, there is a low-affinity high-capacity transporter with a Km for carnitine of 2 to 4 mM [7]. The liver, therefore, is dependent on the plasma level of carnitine to maintain normal tissue concentrations. There is another low-affinity carnitine uptake system in muscle cells [8], which operates at carnitine concentrations between 25 and 200 mM. The myocardial carnitine concentrations are about 40 to 100 times above plasma concentrations (ca. 40 µM).

3.3 Myocardial fatty acid oxidation

The activation of long-chain free fatty acids to their CoA-derivatives represents the first step towards their oxidative degradation. The necessary acyl-CoA synthetase system is present in the heart muscle cell in different localities with different specificities. 80% of its activity is found at the outer mitochondrial membrane and the remainder is connected to the endoplasmatic reticulum [9].

3.4 Carnitine plasma membrane transport defect (systemic carnitine deficiency)

The importance of fatty acids as substrate for myocardial energy metabolism can be indirectly derived from the importance of disturbances of carnitine metabolism for the development of cardiomyopathy. In 1975, primary carnitine deficiency was described by Karpati et al. in an 11-year-old boy [10]. In this patient and in others subsequently described, a hepatocerebral dysfunction (Reye-like syndrome), an increasing muscle weakness and in most cases a dilated cardiomyopathy represented the characteristic clinical features [11]. In some cases endocardial fibroelastosis represented the central cardiological problem [12]. In 8 publications on 24 patients with primary carnitine deficiency, the cardiological problem was distributed in the following manner: 14/24 dilated cardiomyopathy; 7/24 hypertrophic cardiomyopathy and 3/24 endocardial fibroelastosis [11, 12]. Hypertrophic myocytes showed a distinct lipid accumulation [13]. In only 3 of these 24 patients was the myocardial carnitine concentration reported and described as severely decreased. In the cases reported by Eriksson et al. the concentration was even below 1 % of control levels [14]. The most frequently described electrocardiographic changes associated with a carnitine deficiency were ST-T inversions in the left precordial leads. Heterozygotes may also show cardiac symptoms [15]. Patients with primary carnitine deficiency respond very well to carnitine supplementation [12, 16].

An inherited defect in the plasma membrane carnitine transport was first described as primary carnitine deficiency in 1988 [17]. It is defined as a failure of carnitine transport in kidney, muscle including heart muscle, leukocytes and fibroblasts. It is caused by mutations in a gene encoding a sodium ion-dependent carnitine transporter protein termed organic cation transporter N2 (OCTN2) [18]. Many of the described patients presented between 3 months and 2.5 years with episodes characterized by hypoketotic hypoglycaemia, hyperammonaemia, some with cardiomyopathy and/or skeletal muscle weakness. Cardiomyopathy alone was the presenting sign in about 50% of cases. Patients with cardiomyopathy usually presented later (1 to 7 years), and without evidence of hypoglycaemia. In primary carnitine deficiency, the serum carnitine concentrations are always below 10 µM. There was no correlation between residual carnitine uptake and severity of clinical presentation [19].

3.5 The carnitine palmitoyltransferase (CPT) system

Transport of long-chain fatty acids (LCFA) from cytosol into the mitochondrial matrix space requires a special protein association, the carnitine palmitoyltransferase system that reversibly catalyzes the following reaction:

palmitoyl-CoA + carnitine → palmitoylcarnitine + CoA-SH

There are two functionally separate forms of CPT: a CPT I, localized within the outer mitochondrial membrane, which catalyzes the formation of acylcarnitine from carnitine and acyl-CoA and a CPT 2, localized at the inner side of the inner mitochondrial membrane, which catalyzes the formation of acyl-CoA from acylcarnitine and CoA. CPT 1 controls the fatty acid flux through the esterification due to its sensitivity to malonyl-CoA. Malonyl-CoA is the first intermediate in the pathway of fatty acid synthesis and is a potent CPT 1 inhibitor. During fasting, the malonyl-CoA level decreases and CPT 1 is active so that LCFA oxidation and subsequently ketogenesis is enhanced. CPT 1 constitutes an important element for maintenance of energy homeostasis in heart and skeletal muscle. Apart from mitochondria, other subcellular organelles, such as peroxisomes and microsomes, also contain CPT-like enzyme activities.

3.5.1 CPT 1

Two tissue specific CPT 1 isoforms, the so-called "liver" (L) and "muscle" (M) CPT 1, have been shown to exist. L-CPT 1 and CPT 2 share approximately 50 % homology in the major part of their sequences with the exception of their N-termini [20]. They differ with respect to the tightness of their membrane association. While CPT 2 is loosely associated with the inner side of the inner mitochondrial membrane, CPT 1 is embedded within the outer membrane [21, 22]. The CPT 1 protein consists of a single polypeptide containing both inhibitor-binding and catalytic domains. Two L-CPT 1 mutations have been reported. M-CPT 1 predominates in skeletal muscle, heart, adipose tissue and testis [23]. Mitochondrial CPT 1 isoform switching has been established in the developing heart: while L-CPT 1 represents a very minor constituent of the CPT complex in the adult heart [24]. Human L-CPT 1 is expressed in liver, lymphocytes and fibroblasts [25], but not in skeletal muscle [26]. Conversely, CPT 2 is an ubiquitous enzyme in both rat and human [27]. L-CPT 1, M-CPT 1 and CPT 2 genes have been localized to chromosomes 11q13.1-p13.5, 22q13.31-q13.32 and 1p32 respectively [28].

3.5.2 CPT 1 deficiencies

3.5.2.1 Liver-type CPT 1 deficiency

Since the first report in 1981 [29], several patients have been reported [25, 26, 29]. The first presenting symptom either is a Reye-like attack with hypoketotic hypoglycaemia, not constantly associated with hepatomegaly and acute liver failure. Heart involvement is classically generally absent in L-CPT 1 deficiency. However several cases with slight cardiomegaly [26, 30] or heart rhythm disorders [30, 31] have been reported, while a slight myocardial steatosis has been documented in one case [32]. In contrast with other defects of mitochondrial fatty acid oxidation, L-CPT 1 deficiency is almost constantly associated with an elevated level of plasma carnitine [33].

3.5.2.2 Muscle-type CPT 1 deficiency

Hitherto no case of M-CPT 1 deficiency has been reported. A first possibility to explain this fact is that a loss of M-CPT 1 might be incompatible with life, given the importance of this enzyme for cardiac function. If the condition does exist, however, a cardiomyopathy and peripheral myopathy has to be expected in neonates, infants and adults.

3.5.3 CPT 2 deficiencies

There are several presentations of CPT 2 deficiencies which can be individualized according to the age of onset and the tissue distribution of the symptoms. More than 25 mutations have been characterized in patients with the adult, infantile or neonatal form of CPT 2 deficiency. Truncating CPT 2 mutations are associated with the severe neonatal form of the disease.

Adult form

This form is dominated by a muscular symptomatology. Since the first description in 1973 [34], many patients have been reported [30, 35]. 80% of the patients are males although the disease is inherited in an autosomal recessive manner. The first symptoms most often occur between 6 and 20 years of age, but age at onset may be over 50 years or as early as 4 years in some patients. The clinical symptoms usually consist of recurrent attacks of myalgia and muscle stiffness or weakness, occasionally associated with myoglobinuria. The rhabdomyolysis may occasionally be complicated by two kinds of life-threatening events: acute renal failure as a result of myoglobinuria and respiratory insufficiency secondary to respiratory muscle involvement. Symptoms are usually prompted by prolonged exercise and less commonly by fasting, high-fat intake, exposure to cold, mild infection (especially in children), fever, emotional stress, general anaesthesia, or drugs such as diazepam or ibuprofen. The phenotypic expression may be highly variable within a family, with a clinical picture ranging from asymptomatic to lethal [36, 37]. The clinical symptomatology is restricted to skeletal muscle without liver or heart involvement. Biological markers suggestive of CPT 2 deficiency in patients with recurrent rhabdomyolysis are: markedly elevated serum CK and transaminase levels during attacks or after fasting or during prolonged exercise. During intercritical periods these values are usually normal. Carnitine concentrations may be decreased [38, 39] or normal. Serum triglycerides and cholesterol are elevated in about 20% and 10% of patients respectively. Fasting ketogenesis is commonly delayed or decreased. Muscle lipid storage is found in 20% of patients. Other structural anomalies include atrophy or necrosis of type 1 muscle fibre.

Infantile form

Several patients have been reported to have suddenly died, mostly prior to 1 year of age. Age of onset is between 6 months and 2 years, most often prior to 1 year. The clinical picture involves recurrent attacks of acute liver failure with hypoketotic hypoglycaemia, resulting in coma and seizures and transient hepatomegaly. Heart involvement is present in about 50% of cases, occurring either as dilated and hypertrophic cardiomyopathy which may spontaneously recover or as arrhythmias and conduction disorders [40]. Routine laboratory tests commonly show metabolic acidosis and increased levels of ammonia. Low levels of total and free carnitine associated with an increase in the long-chain acylcarnitine fraction are constantly found. Hepatic steatosis is a constant feature. Paroxysmal heart beat disorders are considered to cause sudden death, usually during the 1st year of life as reported in several patients [40].

Neonatal form

The neonatal onset form of CPT 2 deficiency appears to be markedly more severe than the infantile onset form of the disease. Several patients have been reported [41]. Many patients were reported to have suddenly died, most often during the 1st month of life. Malformations like dysmorphic features, cystic renal dysplasia and neuronal migration defects have been reported [42]. The symptom free interval between birth and onset of the acute metabolic symptoms ranges from a few hours to 4 days of life. Respiratory distress, hypoglycaemia with seizures and hepatomegaly and heart involvement presenting as cardiomegaly associated with rhythm and conduction disorders are constant features. Metabolic acidosis and hyperammonaemia are commonly found. It is speculated that increased concentrations of long-chain acylcarnitines may promote cardiac arrhythmia. The severity of disease is related to the residual CPT 2 activity. Muscular patients have a residual activity above 15 % of controls, while it is below 10 % of control values in "hepato-cardio-muscular" patients. The acylcarnitine profile using tandem-mass spectrometry shows a prominent peak of C16 species in infantile/neonatal type CPT 2-deficient patients [43].

3.5.4 Carnitine translocase deficiency

Carnitine/acylcarnitine-translocase is one of 10 carrier proteins required for substrate transfer between the cytosol and the mitochondrial matrix space [44]. This translocase has been isolated and characterized as a 32.5 kDa protein [45]. The translocase activity depends on cardiolipin, a specific phospholipid of the mitochondrial membrane. With increasing chain length of the esterified fatty acid, the activity of carnitine acyltranslocase increases, the lowest affinity therefore being towards free carnitine [46]. Carnitine translocase deficiency was first described in a newborn in 1992 by Stanley et al. [47]. The patient presented with seizures, apnoea and bradycardia at the age of 36 hours after a prolonged fasting period. Repeatedly, premature ventricular contractions as well as ventricular tachycardia were described as the predominant cardiac symptoms. The ECG showed signs of a slight ventricular hypertrophy and ultrasound revealed a decreased ejection fraction. The most important diagnostic hint is the insufficient ketone body formation as well as the constantly increased serum long-chain acylcarnitine concentration. Hereby is demonstrated on the one hand the inability of the long-chain fatty acid transport into the mitochondrial matrix space and on the other hand the defect is localized beyond the CPT 1 reaction. Only deficiencies of carnitine acyltranslocase and CPT 2 remain as diagnostic possibilities. Translocase deficiency can therefore be added to the carrier protein defects of the mitochondrial membrane and in this aspect is akin to hyperornithinaemia and the HHH syndrome [48]. The decreased serum carnitine concentrations in patients with translocase deficiency are indicative of an additional secondary carnitine deficiency, which can be explained by the inhibitory action of long-chain acylcarnitines on the cytoplasmatic carnitine transport [49]. At the same time the toxic effect of long-chain acyl carnitine concentrations is accepted to be the cause of cardiac arrhythmias. As a very severe form of arrhythmia in translocase deficiency, complete atrioventricular block was described [50].

3.5.5 Secondary carnitine deficiency

Secondary carnitine deficiencies are known to occur in inborn defects of intermediary metabolism, in particular of amino acids (organic acidaemias). They result in a decreased availability of free carnitine and may represent the basis of functional myocardial disturbances. The occurrence of cardiac dysfunction as a consequence of secondary carnitine deficiency caused by valproic acid therapy, responding to carnitine supplementation, represents an important insight into pathophysiology [51].

References

[1] Tripp, M.E.: Developmental cardiac metabolism in health and diesese. Pediatr.Cardiol. **10**, 150-158 (1989).
[2] Bing, R.J., A. Siegel, A. Vitale: Metabolic studies on human heart in vivo; studies on carbohydrate metabolism of human heart. Amer.J.Med. **15**, 284-296 (1953).
[3] Bing, R.J.: Myocardial metabolism. Circulation **12**, 635-647 (1955).
[4] Drake, A.J.: Substrate utilization in the myocardium. Basic Res. Cardiol. **77**, 1-11 (1982).
[5] Opie, L.H.: Metabolism of the heart in health and disease. Amer.J.Cardiol. **77**, 100-122 (1969)
[6] Bremer, J.: Carnitine-metabolism and functions. Physiol. Rev. **63**, 1420 (1983).
[7] Sandor, A., G. Krispal, B. Melegh, I. Alkonyi: Release of carnitine from the perfused rat liver. Biochim.Biophys.Acta **835**, 83 (1985).
[8] Martinuzzi, A., L. Vergani, M. Rosa, C. Angelici: L-Carnitine uptake in differentiating human cultured muscle. Biochim.Biophys.Acta **1095**, 217 (1991).
[9] DeJong, J.W., W.C. Hülsmann: A comparative study of palmitoyl-CoA synthetase activity in rat liver, heart and gut mitochondrial and microsomal preparations. Biochim.Biophys.Acta **197**, 127-135 (1970).
[10] Karpati, G., S. Carpenter, A.G. Engel, G. Watters, J. Allen, S. Rothman, G. Klassen, O.A. Mamer: The syndrome of systemic carnitine deficiency. Neurology **25**, 16-24 (1975).
[11] Chapoy, P.R., C. Angelini, W.J. Brown, J.E. Stiff, A.L. Shug, S.D. Cederbaum: Systemic carnitine deficiency. A treatable inherited lipid-storage disease presenting as Reye's syndrome. New Engl.J.Med. **303**, 1389-1394 (1980).
[12] Tripp, M.E., M.L. Katcher, H.A. Peters, E.F. Gilbert, S. Arya, R.J. Hodach, A.L. Shug: Systemic carnitine deficiency presenting as familial endocardial fibroelastosis. A treatable cardiomyopathy. New Engl. J.Med. **305**, 385-390 (1981).
[13] Ino, T., W.G. Sherwood, L.N. Benson, G.J. Wilson, R.M. Freedom, R.D. Rowe: Cardiac manifestations in disorders of fat and carnitine metabolism in infancy. J.Amer.Coll.Cardiol. **11**, 1301-1308 (1988).
[14] Eriksson, B.O., S. Lindstedt, I. Nordin: Hereditary defect in carnitine membrane transport is expressed in skin fibroblasts. Eur.J.Pediatr. **147**, 662-663 (1988).
[15] Garavaglia, B., G. Uziel, F. Dworzak, F. Carrara, S. DiDonato: Primary carnitine deficiency heterozygote and intrafamilial phenotypic variation. Neurology **41**, 1691-1693 (1991).
[16] Waber, L.J., D. Valle, C. Neill, S. DiMauro, A. Shug: Carnitine deficiency presenting as familial cardiomyopathy: A treatable defect in carnitine transport. J. Pediatr. **101**, 700.705 (1982).
[17] Treem, W.R., C.A. Stanley, D.N. Finegold, D.E. Hale, P.M. Coates: Primary carnitine deficiency due to a failure of carnitine transport in kidney, muscle, and fibroblasts. New Engl.J.Med **319**, 1331 (1988).
[18] Nezu, J., I. Tamai, A. Oku, R. Ohashi, H. Yabuuchi, N. Hashimoto, H. Nikaido, Y. Sai, A. Koizumi, Y. Shoji, G. Takada, T. Matsuishi, M. Yoshino, H. Kato, T. Ohura, G. Tsujimoto, J. Hayakawa, M. Shimane, A. Tsuji: Primary systemic carnitine deficiency is caused by mutations in a gene encoding sodium ion-dependent carnitine transporter. Nat. Genet. **21**, 91-94 (1999).
[19] Lamhonwah, A.M., S. E. Olpin, R.J. Pollitt, C. Vianey-Saban, P. Divry, N. Guffon, G.T. Besley, R. Onizuka, L.J. Meirleir, L. Cvitanovic-Sojat, I. Baric, C. Dionisi-Vici, K. Fumic, M. Maradin, I. Tein: Novel OCTN2 mutations: no genotype-phenotype correlations: early carnitine therapy prevents cardiomyopathy. Am.J.Med.Genet. **111**, 271-284 (2002).

[20] Kolodziej, M.P., V.A. Zammit: Mature carnitine palmitoyltransferase I retains the N-terminus of the nascent protein in rat liver. FEBS Lett. **327**, 294-296 (1993)
[21] Woeltje, K.F., V. Esser, B.C. Weis, A. Sen, W.F. Cox, M.J. McPhaul, C.A. Slaughter, D.W. Foster, J.D. McGarry: Cloning, sequencing and expression of a cDNA encoding rat liver mitochondrial carnitine palmitoyltransferase II. J.Biol.Chem. **265**, 10720-10725 (1990).
[22] Cohen, I., C. Kohl, J.D. McGarry, J. Girard, C. Prip-Buus: The N-terminal domain of rat liver carnitine palmitoyltransferase I mediates import into the outer mitochondrial membrane and is essential for activity and malonyl-CoA sensitivity. J.Biol.Chem. **273**, 29896-29904 (1998).
[23] Esser, V., N.F. Brown, A.T. Cowan, D.W. Foster, J.D. McGarry: Expression of a cDNA isolated from rat brown adipose tissue and heart identifies the product as the muscle isoform of carnitine palmitoyltransferase 1 (M-CPT 1): M-CPT 1 is the predominant CPT 1 isoform expressed in both white (epididymal) and brown adipocytes. J.Biol.Chem. **271**, 6972-6977 (1996).
[24] Brown, N.F., B.C. Weis, J.E. Husti, D.W. Foster, J.D. McGarry: Mitochondrial carnitine palmitoyltransferase I isoform switching in the developing rat heart. J.Biol.Chem. **270**, 8952-8957 (1995).
[25] Demaugre, F., J.P. Bonnefont, G. Mitchell, N. Nguyen-Hoang, A. Pelet, M. Rimoldi, S. DiDonato, J.M. Saudubray: Hepatic and muscular presentations of carnitine palmitoyltransferase deficiency: Two distinct entities. Pediatr.Res. **24**, 308-311 (1988).
[26] Tein, I., F. Demaugre, J.P. Bonnefont, J.M. Saudubray: Normal muscle CPT I and CPT II activities in hepatic-presentation patients with CPT I deficiency in fibroblasts. Tissue specific isoforms of CPT I ? J.Neurol.Sci. **92**, 229-245 (1989).
[27] Woeltje, K.F., V. Esser, B.C. Weis, A. Sen, W.F. Cox, J.G. Schroeder, S.T. Liao, D.W. Foster, J.D. McGarry: Inter-tissue and inter-species characteristics of the mitochondrial carnitine palmitoyltransferase enzyme system. J.Biol.Chem. **265**, 10714-10719 (1990).
[28] Britton, C.H., D.W. Mackey, V. Esser, D.W. Foster, D.K. Burns, D.P. Yarnall, P. Froguel, J.D. McGarry: Fine chromosome mapping of the genes for human liver and muscle carnitine palmitoyltransferase I. Genomics **40**, 209-211 (1997).
[29] Bougneres, P.F., J.M. Saudubray, C. Marsac, O. Bernard, M. Odièvre, J. Girard: Fasting hypoglycaemia resulting from hepatic deficiency. J.Pediatr **98**, 742-746 (1981)
[30] Schaefer,J., S. Jackson, F. Taroni, P. Swift, D.M. Turnbull: Characterisation of carnitine palmitoyltransferases in patients with carnitine palmitoyltransferase deficiency: Implications for diagnosis and therapy. J.Neurol.Neurosurg.Psychiat. **62**, 169-176 (1997)
[31] Bergman, A.J.I.W., A.M.G. Donckerwolcke, M. Duran, J.A.M. Smeitink, M. Mousson, C. Vianey-Saban, B.T. Poll-The: Rate-dependent distal renal tubular acidosis and carnitine palmitoyltransferase I deficiency. Pediatr.Res. **5**, 582-588 (1994).
[32] Vianey-Saban, C., B. Mousson, D. Floret, C. Bertrand, R. Dumoulin, M.T. Zabot, P. Divry: Carnitine palmitoyltransferase I deficiency presenting as a Reye-like syndrome without hypoglycaemia. Eur.J.Pediatr. **152**, 334-338 (1993).
[33] Stanley, C.A., F. Sunaryo, D.E. Hale, J.P. Bonnefont, F. Demaugre, J.M. Saudubray: Elevated plasma carnitine in the hepatic form of carnitine palmitoyltransferase l deficiency. J.Inher.Metab.Dis. **15**, 785-789 (1992)
[34] Di Mauro, S., P.M. Di Mauro: Muscle carnitine palmitoyltransferase deficiency and myoglobinuria. Science **182**, 929-931 (1973)
[35] Bank, W.J., S. DiMauro, E. Bonilla, D.M. Capuzzi, L.F. Rowland: A disorder of muscle lipid metabolism and myoglobinuria. Absence of carnitine palmityl transferase. New Engl.J.Med. **292**, 443-449 (1975).
[36] Kelly, K.S., J.S. Garland, T.T. Targ, A.L. Shug, M.J. Chusid: Fatal rhabdomyolysis following influenza infection in a girl with familial carnitine palmityltransferase deficiency. Pediatrics **84**, 312-316 (1989).
[37] Schröder, J.P., W. Mau, S. Schumacher, S. Zierz: Abnorme Regulation der Carnitin-palmitoyltransferase bei eineiigen Zwillingen als Ursache einer Rhabdomyolyse. Dtsch.Med.Wochenschr. **115**, 337-341 (1990).
[38] Iosanescu, V., G. Hug, C. Hoppel: Combined partial deficiency of muscle carnitine palmitoyl transferase and carnitine with autosomal dominant inheritance. J.Neurol.Neurosurg.Psychiat. **43**, 679-682 (1980).
[39] Heier, M.S., P. Dietrichson, S. Landaas: Familial combined deficiency of muscle carnitine and carnitine palmityltransferase (CPT). Acta Neurol.Scand. **74**, 479-485 (1986).
[40] Demaugre, F., J.P. Bonnenfont, M. Colonna, C. Cepanec, J.P. Leroux, J.M. Saudubray : Infantile form of carnitine palmitoyltransferase II deficiency with hepatomuscular symptoms and sudden death. Physiolopathological approach to carnitine palmitoyltransferase II deficiencies. J.Clin.Invest. **87**, 859-864 (1991).
[41] Hug, G., K.E. Bove, S. Soukup: Lethal neonatal multiorgan deficiency of carnitine palmitoyltransferase II. New Engl.J.Med. **325**, 1862-1864 (1991)
[42] North, K., C.L. Hoppel, U. De Girolami, H. Kozakewich, M. Korson: Lethal neonatal deficiency of carnitine palmitoyltransferase 2 associated with dysgenesis of the brain and kidneys. J.Pediatr. **127**, 414-420 (1995).
[43] Millington, D.S., N. Terada, D.H. Chace, Y.T. Chen, J.H. Ding, N. Kodo, C.R. Roe: The role of tandem-mass spectrometry in the diagnosis of acid oxidation disorders. Prog.Clin.Biol.Res. **375**, 339-354 (1992).
[44] Kramer, R., F. Palmieri: Molecular aspects of isolated and reconstituted carrier proteins from animal mitochondria. Biochim.Biophys.Acta **974**, 1-23 (1989).
[45] Indivieri, C., A. Tonazzi, G. Prezioso, F. Calmieri: Kinetic characterization of the reconstituted carnitine carrier from rat liver mitochondria. Biochim. Biophys.Acta **1020**, 81-86 (1991).
[46] Indivieri, C., A. Tonazzi, G. Prezioso, F. Calmieri: Kinetic characterization of the reconstituted carnitine carrier from rat liver mitochondria. Biochim.Biophys.Acta **1065**, 231-238 (1991).
[47] Stanley, C.A., D.E. Hale, G.T. Berry, S. Deleeuw, Boxer, J.P. Bonnefont: Brief report: A deficiency of carnitine-acylcarnitine translocase in the inner mitochondrial membrane. New Engl.J.Med. **327**, 19-23 (1992).
[48] Shih, V.E., R. Mandell, A. Herzfeld: Defective orni-

thine metabolism in cultured skin fibroblasts from patients with the syndrome of hyperornithinemia, hyperammonemia and homocitrullinemia. Clin.Chim.Acta **118**, 149-157 (1982).

[49] Stanley, C.A., S. DeLeeuw, P.M. Coates: Chronic cardiomyopathy and weakness or acute coma in children with a defect in carnitine uptake. Ann.Neurol. **30**, 709-716 (1991).

[50] Pande, S.V., M. Brivet, A. Slama, F. Demaugre, C. Aufrant, J.M. Saudubray: Carnitine-acylcarnitine translocase deficiency with severe hypoglycemia and auriculo ventricular block. Translocase assay in permeabilized fibroblasts. J.Clin.Invest. **91**, 1247-1252 (1993).

[51] Bratton, S.L., A.L. Garden, T.P. Bohan, J.W. French, W.R. Clarke: A child with valproic acid-associated carnitine deficiency and carnitine-responsive cardiac dysfunction. J.Child.Neurol. **7**, 414-416 (1992).

4 Defects in long-chain fatty acid oxidation as a cause of cardiomyopathy

H. Przyrembel, A. C. Sewell

The mammalian heart is capable of oxidising carbohydrates, fats and aminoacids, whereby fatty acid oxidation plays the dominant role by inhibiting the oxidation of carbohydrates. Fatty acid oxidation can be inhibited by the presence of glucose, lactate and pyruvate, e.g. during the transition from fasting to the post prandial state [4]. The level of circulating free fatty acids determines the myocardial uptake of these compounds which are rapidly oxidised. The resulting high levels of ATP and acetyl-CoA inhibit the activity of pyruvate dehydrogenase and thus limit the entry of glucose and lactate into the citric acid cycle. At the same time, phosphofructokinase activity is limited by the high citrate levels leading to increased intracellular glucose-6-phosphate which inhibits further glucose uptake [26]. Since fatty acid oxidation is so important for cardiac metabolism it is not surprising that fatty acid oxidation defects manifest with cardiac problems, nevertheless, since these defects are not very common, it appears that some compensation for fatty acid oxidation defects may be possible.

Before discussing the known defects of fatty acid oxidation (Table 1), some description of normal fatty acid oxidation is necessary. In describing the defects, attention will be paid primarily to defects of long-chain acyl-CoA dehydrogenase (LCAD) and long-chain 3-hydroxyacyl-CoA dehydrogenase (LCHAD), since they appear to be the most common. Acylcarnitine translocase deficieney is discussed in the chapter on „carnitine and cardiac energy metabolism".

The major path of fatty acid oxidation is depicted in Figure 1. The cellular uptake of circulating free fatty acids is mediated via fatty acid binding proteins of the plasma membrane. These proteins

Tab. 1: Defects of mitochondrial long-chain fatty acid oxidation.

- Very long-chain acyl-CoA dehydrogenase deficiency
- Long-chain acyl-CoA dehydrogenase deficiency
- Long-chain 3-hydroxyacyl-CoA dehydrogenase deficiency
- (Long-chain 3-ketothiolase deficiency?)
- Mitochondrial trifunctional protein deficiency
- Acylcarnitine translocase deficiency
- (2,4-Dienoyl-CoA reductase deficiency?)
- Multiple acyl-CoA dehydrogenase deficiency
  - ETF deficiency
  - ETF-ubiquinone reductase deficiency
  - Riboflavin responsive form

may be organ specific but are not identical to cytoplasmic fatty acid binding proteins [19, 21]. Uptake by concentration dependent diffusion is also possible [17]. The exact function of the different cytoplasmic fatty acid binding proteins in the control of cytoplasmic, microsomal and mitochondrial enzymatic processes, e.g. the protection of enzymes and cellular structures from the detergent effect of fatty acids, remains hypothetical [20]. It is quite possible that abnormalities in fatty acid binding proteins may lead to defects in fatty acid oxidation, either in the whole organism or in individual organs. The role of the carnitine shuttle after the activation of long-chain fatty acids via acyl-CoA synthetase has already been discussed in a preceding chapter. Short- and medium-chain fatty acids may enter the mitochondrion without prior activation where they are converted to CoA thioesters. Long-chain unsaturated (C16C20) and saturated (C10-C18) fatty acids are converted to their respective acyl-CoA esters by acyl-CoA synthetase situated on the endoplasmic reticulum, microsomal membrane and on the outer mitochondrial membrane. Specific acyl-CoA synthetases exist for very-long-chain fatty acids [17]. Fatty acid CoA esters enter

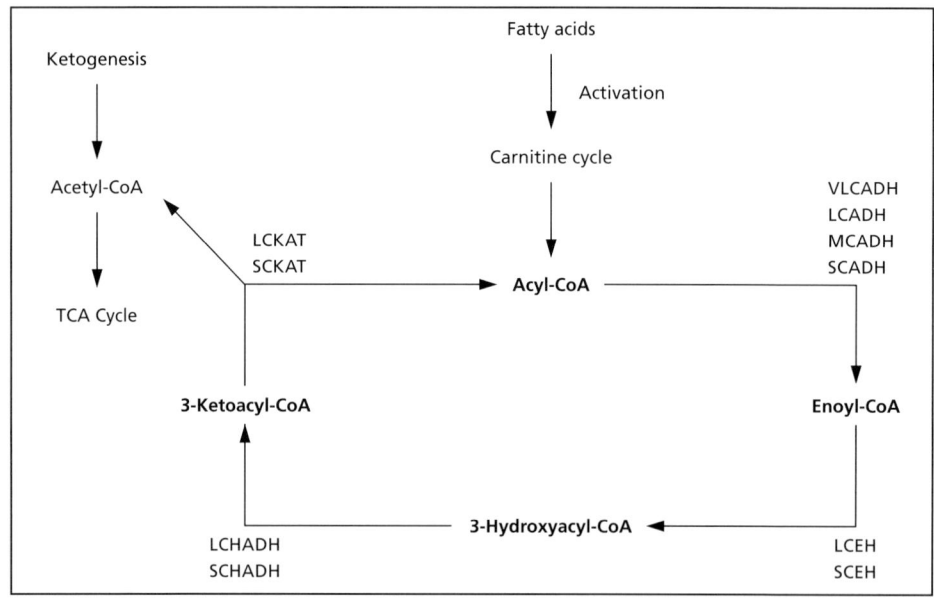

Fig. 1: Mitochondrial Fatty Acid Oxidation.
VLCADH: very long-chain acyl-CoA dehydrogenase
LCADH: long-chain acyl-CoA dehydrogenase
MCADH: medium-chain acyl-CoA dehydrogenase
SCADH: short-chain acyl-CoA dehydrogenase
LCEH: long-chain enoyl hydratase
SCEH: short-chain enoyl hydratase
LCHADH: long-chain 3-hydroxyacyl-CoA dehydrogenase
SCHADH: short-chain 3-hydroxyacyl-CoA dehydrogenase
LCKAT: long-chain keto-acyl thiolase
SCKAT: short-chain keto-acyl thiolase

the mitochondrial β-oxidation spiral which encompasses four reactions leading to the shortening of the molecule by two carbon atoms with the release of free acetyl-CoA. In the first step (Fig. 1), 2-trans-enoyl-CoA results from the action of the chain-length specific acyl-CoA dehydrogenases. In addition to the three known acyl-CoA dehydrogenases (short-chain: C4-C6; medium-chain: C10-C14 with maximum activity for C8-C14; longchain: C10-C20 with maximum activity for C10), a very-long-chain acyl-CoA dehydrogenase specific for C15-C20 has been described [12]. These acyl-CoA dehydrogenases exhibit overlapping activities with substrates of differing chain length. On dehydrogenation of the fatty acid, the covalently bound enzyme co-factor flavine adenine dinucleotide (FAD) is reduced. Cofactor reoxidation requires electron-transfer flavoprotein (ETF) which is reoxidised via ETF: ubiquinone oxidoreductase situated on the inner mitochondrial membrane. This is the point of coupling to the respiratory chain.

The second step is the rehydration of enoyl-CoA to L-3-hydroxyacyl-CoA by two enoyl-CoA hydratases. One hydratase, present in the mitochondrial matrix, catalyses substrates of chain length C4-C16, although with decreasing efficiency. The other, a membrane bound enzyme, exhibits higher activity with longer chain length substrates (maximum C8) and is part of the trifunctional enzyme complex (3-hydroxyacyl CoA dehydrogenase; 3-ketothiolase; enoyl-CoA hydratase) [5].

The third step is a dehydrogenation to 3-keto acyl-CoA via one of two L-3-hydroxyacyl CoA dehydrogenases. The fourth and final step is the release of acetyl-CoA by one of two 3-ketothiolases. The short-chain acetoacetyl-CoA thiolase may be solely active for ketone body metabolism. It remains unknown whether the 3-ketothiolase activity of the trifunctional protein is identical with the long-chain 3-ketothiolase which is active with substrates of all chain lengths. The two carbon atom shortened fatty acid then re-enters the spiral to be further metabolised.

Figure 2 depicts the relationship between fatty acid oxidation and ketone body production with the respiratory chain and citric acid cycle. Defects in fatty acid oxidation - at whatever level - must lead to impaired ketone body production, a lowered energy production via the respiratory chain and diminished gluconeogenesis by the inhibition of pyruvate carboxylase. Thus the most important indication for a fatty acid oxidation defect is hypoketotic hypoglycaemia, particularly upon fasting, or in situations where a high energy requirement cannot be maintained (e.g. infections). The previous discussion has been restricted to the oxidation of saturated fatty acids. Unsaturated fatty acids with one or more double bonds undergo β-oxidation, but require two additional enzymatic steps, namely a cis-trans-enoyl-CoA isomerase and a NADPH dependent 2,4-dienyol-CoA reductase. These are necessary to avoid inhibition of the normal β-oxidation pathway by double bond containing compounds [17].

Abnormalities of mitochondrial ß-oxidation lead to the formation of dicarboxylic acids. This phenomenon is found during ketosis and MCT nutrition and in animals with riboflavin deficiency. An increase in cytoplasmic monocarboxylic acids leads to microsomal omega or omega-1 oxidation. Methyl groups of C10-C12 fatty acids are oxidised by mono-oxygenases to hydroxyl

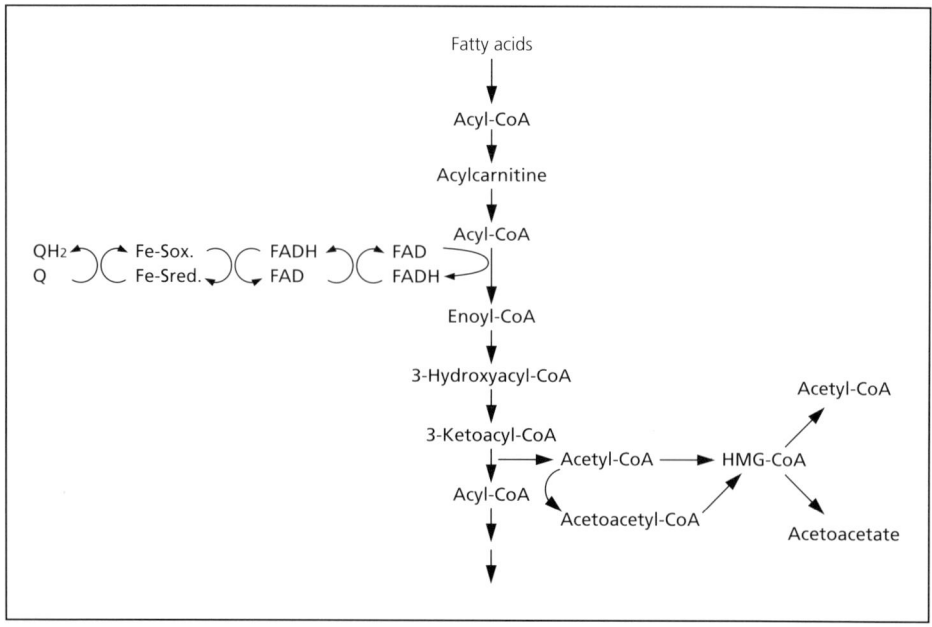

Fig. 2: **Relationship between fatty acid oxidation, ketone body production and the respiratory chain.**

and carbonyl groups. The resulting dicarboxylic acids undergo β-oxidation in both mitochondria and peroxisomes. In peroxisomes, this only proceeds to adipic acid, as long as β-oxidation remains intact. Intermediary metabolites of deranged mitochondrial β-oxidation such as 3-hydroxy fatty acids in the case of 3-hydroxyacyl-CoA dehydrogenase deficiency, can leave the mitochondrion, possibly bound to carnitine, and be oxidized to 3-hydroxydicarboxylic acids which are excreted in the urine. Furthermore, increased mitochondrial fatty acid metabolites may be bound to carnitine via the carnitine acyl transferases and can exit the mitochondrion as carnitine esters.

The most important diagnostic parameters (Table 2) are the simultaneous determination of blood glucose, free fatty acids and ketone bodies, total, free and esterified carnitine in plasma and the urinary excretion of organic acids (dicarboxylic acids and 3-hydroxyacids) and carnitine and carnitine esters. These metabolites are of diagnostic significance at the time of acute decompensation, i.e. in emergency situations provoked by fasting or during controlled loading tests. Confirmation should be obtained by enzyme analysis in cultured skin fibroblasts, tissue biopsies or in leucocytes.

The individual disorders will now be discussed in the order of their first description.

4.1 Long-chain acyl-CoA dehydrogenase (LCAD) deficiency

The first patients were described in 1980 by Naylor et al. [14] and at present fewer than 20 patients are known. Disease progression is variable. Besides very early presentation with cardiac manifestation and high mortality, cases with early acute onset without cardiac problems are known together with rarer cases of later onset. These late onset cases presented with muscular hypotonia, myalgia and myoglobinuria together with elevated creatine kinase levels during fasting, excessive exercise or viral infections.

Typical onset is in the early neonatal period (31 h in one patient) with progressive obtundation resulting in coma, hypotonia, food refusal, hepatomegaly and cardiomegaly which in one case coincided with pericardial effusion [22]. Many patients die during the first or subsequent attacks. Hypoketotic hypoglycaemia is usually present, resulting in an increased ratio of free fatty acids to 3-hydroxybutyric acid. Total carnitine is normal or decreased with ca. 90% being esterified (normal = < 25%). Plasma ammonia and lactate levels can be elevated. A urinary dicarboxylic aciduria (saturated and unsaturated) of C6-C14 chain length is present without glycine conjugates. Carnitine esters (C10-C12 chain length) are also present. MCT loading results in a normal ketone body production. Liver pathology reveals both macro- and macro-vesicular fat deposits together with unspecific mitochondrial abnormalities such as increased intercristal spaces, enlarged matrix and crystalloid inclusions.

Treatment consists of avoidance of long fasting periods which may necessitate nocturnal gastric tube feeding. Nutrition should be carbohydrate rich and fat reduced but can contain MCT. Carnitine treatment may be beneficial. Some patients treated with this regime have demonstrated a normalisation of the cardiomegaly, but the muscular hypotonia has persisted. The urinary excretion pattern of patients on treatment can be completely normal. Enzymatic confirmation is possible. The LCAD activity in parents of patients is slightly decreased (37-67%) [8, 9, 22]. The enzyme defect appears to be due to a point mutation in all cases.

Tab. 2: Diagnostic investigations in cases of suspected long-chain fatty acid oxidation defects.

- **Acute Situation**
  a) Blood: glucose
          free fatty acids
          ketone bodies
          carnitine (free and esterified)
          lactate
          ammonia
          creatine kinase
  b) Urine: organic acids
          carnitine esters
          glycine conjugates
- **Fasting Test (cave!):** as above
- **Fat loading (MCT/LCT):** as above
- **Phenylpropionate loading test:**
  presence of phenylhydracrylic acid

4.2 3-Hydroxyacyl-CoA dehydrogenase (LCHAD) deficiency

This disorder was first described in 1988 by Kelley and Morton [11] in a 4.5

month old girl who presented with hepatopathy, hepatomegaly, hypoglycaemia and coma. Urinary analysis revealed a saturated and unsaturated dicarboxylic aciduria with increased 3-hydroxyadipic, 3-hydroxysuberic, 3-hydroxysebacic and 3-hydroxyoctanoic acids. Treatment with MCT increased this pattern. There was a fasting intolerance with hypoketotic hypoglycaemia and a lactic acidaemia (which may be quite impressive: >7 mmol/L) [2, 7, 10, 18]. Free carnitine in plasma was decreased with increased esterified carnitine, particularly long-chain carnitine [18]. Biventricular cardiohypertrophy was present. Microscopy of a liver biopsy revealed evidence of cirrhosis with lipid storage and electron microscopy showed unspecific mitochondrial changes. The enzyme defect was first delineated by Wanders et al. [23] in a patient who was successfully treated with MCT and who excreted phenylacrylic acid on loading with phenylpropionic acid. Patients can present in the neonatal period or after the 1st year of life; nevertheless, the clinical picture is always similar. Certain patients had peripheral neuropathy [2, 6], pericardial effusion [10], retinal pigment degeneration [2] and myoglobinuria [6]. Treatment consists of avoidance of fasting, a fat reduced and carbohydrate rich diet with or without the inclusion of MCT or carnitine. This regime has led to normal growth and development in very few patients [7, 13], whereas the majority have died, usually before therapeutic intervention can be attempted [18]. A benign case has been described [15] who presented at the age of 6 months with developmental delay, feeding difficulties and hypotonia. She was treated with diet without MCT but with carnitine and at 10 years of age, except for provocational intolerance and possible retinal changes, was completely normal.

Confirmation by enzyme analysis in cultured skin fibroblasts is possible. Prenatal diagnosis in chorionic villus tissue is also available. Preliminary results suggest that a single mutation may be responsible for this disease.

An interesting observation has been made in that mothers of children with LCHAD deficiency have had a fatty liver or HELLP (haemolysis, increased liver enzymes, low platelets) syndrome during pregnancy [25].

4.3 2,4-Dienoyl reductase deficiency

Until now only one patient with this disorder has been described [16]. This patient, a dystrophic full-term neonate, presented at 2 days of age with sepsis, drinking problems and hypotonia. Two months later, symptoms included microcephaly with enlarged ventrieles and dysproportion was present. Plasma lysine was elevated (as yet unclear). Total and free plasma carnitine were decreased whereas esterified carnitine was increased. Abnormal carnitine esters, in particular 2-trans-4-cis-decadienoylcarnitine, were present in urine. The activity of 2,4-dienoyl-coenzyme A reductase activity with 2-trans-4-cis-dienoyl-CoA as substrate was 17% of normal in muscle and 40% of normal in liver. The patient died at 4 months of age as a result of cardiac insufficiency due to biventricular hypertrophy despite treatment with MCT diet and carnitine.

4.4 Trifunctional protein deficiency

A single patient with this deficiency has been described [24]. However, a case of LCHAD deficiency revealed a decreased activity of the 3-ketothiolase to 40% of controls [2]. Nevertheless, seven of eight cases of LCHAD deficiency showed an isolated enzyme defect and one case only exhibited a combined deficiency of long-chain enoyl-CoA hydratase, 3-hydroxyacyl-CoA dehydrogenase and 3-ketoacyl-CoA thiolase. The patient presented at 48 h of life with hypoglycaemia and hypotonia. Artificial ventilation was necessary and the child died aged 30 days due to acute cardiac insufficiency (hypokinetic dilated cardiomyopathy). Urinary analysis revealed a dicarboxylic aciduria (C6-C10), including unsaturated derivatives, together with 3-hydroxyadipic, 3-hydroxysuberic, 3-hydroxysebacic and 3-hydroxydodecanedioic acids.

4.5 Very-long-chain acyl-CoA dehyrogenase (VLCAD) deficiency

A patient with this defect was described by Bertrand et al. [3]. A previous sibling had died at 2 days of age and revealed a massive hepatic fat deposition. The index patient refused feeds at 2 days of age, exhibited ventricular fibrillation and had a respiratory arrest. A metabolic acidosis was present with elevated creatine kinase activity and a massive urinary dicarboxylic aciduria (C6 > C8 > C10 > C12). Free carnitine in plasma was decreased. A phenylpropionic acid loading test was normal. The patient was successfully treated with fat restricted diet and carnitine and by 2 years of age had developed normally with normal cardiac function. Palmitate oxidation in fibroblasts was decreased whereas that of octanoate was normal. Long-chain acyl-CoA dehydrogenase activity in the supernatant of homogenised fibroblasts was normal. The activity of the membrane-bound very-long-chain acyl-CoA dehydrogenase with palmitate as substrate in pellets of fibroblast homogenates was less than 5% of controls.

4.6 Long-chain 3-ketothiolase deficiency

Bennett and Sherwood [1] described three possible cases of long-chain 3-ketothiolase deficiency. The first case died aged 23 months due to rapid progressing liver failure and chronic developmental motor retardation. A fatty liver was revealed on autopsy. The second case was investigated at 1 year of age because of failure to thrive, weight loss, hypotonia and muscle weakness. The third patient presented at 9 months with muscle weakness, vomiting and weight loss. All three patients (the second and third during a 21 h fasting test) showed the same metabolites in urine: saturated and unsaturated dicarboxylic acids (C6-C12), 3-hydroxydicarboxylic acids (C6-C12) including unsaturated derivatives (C10:1, C12:1) and, in particular, 3-ketodicarboxylic acids (C6-C12). The origin of these metabolites could be a defect of the long-chain 3-ketoacyl-CoA-thiolase or possibly of the trifunctional protein.

Although cardiomyopathy may present in patients with multiple acyl-CoA dehydrogenase deficiency, this disorder will not be discussed here. In this disorder, not only is the oxidation of long-chain fatty acids disturbed, but also that of short-chain and medium-chain fatty acids. The oxidation of branched-chain fatty acids stemming from the metabolism of amino acids, glutaric acid and sarcosine is also affected. The defect in ETF or ETF-ubiquinone oxidoreductase leads to a situation akin to that of fatty acid oxidation disorders; however, the urinary organic acid excretion pattern in this disorder is far more complicated and much more specific.

Establishing a specific diagnosis of a defect in the oxidation of long-chain fatty acids is complicated by the clinical presentation and unspecific laboratory results. It is most important to recognise the general clinical features, i.e. acute episodes, usually triggered by infections, gastro-enteritis or long fasting periods, with coma, convulsions, hypotonia, cardiomegaly and hepatomegaly. Chronic symptoms of developmental motor retardation with hypotonia and muscle weakness, also with hepatopathy and cardiomyopathy, should be borne in mind. Unusual features such as neuropathy, retinal pigment degeneration and stress induced myoglobinuria deserve particular attention.

Since the laboratory investigation of such defects plays a crucial role in delineating defects of long-chain fatty acid oxidation, it can never be too strongly emphasised how important it is to obtain the relevant blood and urine samples during the acute phase of the disease in order to avoid hazardous fasting tests.

References

[1] Bennett, M. J., W. G. Sherwood: 3-Hydroxydicarboxylic and 3-ketodicarboxylic aciduria in three patients: evidence for a new defect in fatty acid oxidation at the level of 3-ketoacyl-CoA thiolase. Clin. Chem. **39**, 897-901 (1993).
[2] Bertini, E., C Dionisi-Vici, B. Garavaglia, A. B. Burlina, M. Sabatelli, M. Rimoldi, A. Bartuli, G. Sabetta, S. DiDonato: Peripheral sensory-motor polyneuropathy, pigmentary retinopathy and fatal cardiomyopathy in long-chain 3-hydroxy-acyl-CoA dehydrogenase deficiency. Eur. J. Pediatr. **151**, 121-126 (1992).
[3] Bertrand, C., C. Largilliere, M.-T. Zabot, M. Mathieu, C. Vianey-Saban: Very long chain acyl-CoA dehydrogenase deficiency: identification of a new inborn error of mitochondrial fatty acid oxidation in fibroblasts. Biochim. Biophys. Acta **1180**, 327-329 (1993).
[4] Brosnan, J. T., K. Reid: Inhibition of palmitoyil-carnitine oxidation by pyruvate in rat heart mitochondria. Metabolism **34**, 588-593 (1985).
[5] Carpenter, K., R. J. Pollitt, B. Middleton: Human liver long-chain 3-hydroxyacyl-coenzyme A dehydrogenase is a multifunctional membrane-bound beta-oxidation enzyme of mitochondria. Biochem. Biophys. Res. Commun. **183**, 443-448 (1992).
[6] Dionisi-Vici, C., A. B. Burlina, E. Bertini, C. Bachmann, M. R. M. Mazziotta, E. Zacchello, G. Sabetta, D. E. Hale: Progressive neuropathy and recurrent myoglobinuria in a child with long-chain 3-hydroxyacyl-coenzyme A dehydrogenase deficiency. J. Pediatr. **118**, 744-746 (1991).
[7] Duran, M., R. J. A. Wanders, J. P. de Jager, L. Dorland, L. Bruinvis, D. Ketting, L. Ijlst, F. J. van Sprang. 3-Hydroxydicarboxylic aciduria due to long-chain 3-hydroxyacyl-coenzyme A dehydrogenase deficiency associated with sudden neonatal death: protective effect of medium-chain triglyceride treatment. Eur. J. Pediatr. **150**, 190-195 (1991).
[8] Hale, D. E., C. A. Stanley, P. M. Coates: The longchain acyl-CoA dehydrogenase deficiency. In: Fatty Acid Oxidation. Clinical, Biochemical and Molecular Aspects. K. Tanaka, P. M. Coates (eds.) Alan R. Liss, Inc., New York, pp. 303-311 (1990).
[9] Indo, Y., P. M. Coates, D. E. Hales, K. Tanaka: Immunochemical characterization of variant long-chain acyl-CoA dehydrogenase in cultured fibroblasts from nine patients with long-chain acyl-CoA dehydrogenase deficiency. Pediatr. Res. **30**, 211-215 (1991).
[10] Jackson, S., K. Bartlett, J. Land, E. R. Moxon, R. J. Pollitt, J. V. Leonard, D. M. Turnbull: Longchain 3-hydroxyacyl-CoA dehydrogenase deficiency. Pediatr. Res. **29**, 406-411 (1991).
[11] Kelley, R. L., D. H. Morton: 3-Hydroxyoctanoic aciduria: Identification of a new organic acid in the urine of a patient with non-ketotic hypoglycemia. Clin. Chim. Acta **175**, 19-26 (1988).
[12] Kelley, R. L.: Beta-oxidation of long-chain fatty acids by human fibroblasts: evidence for a novel long-chain acyl-coenzyme A dehydrogenase. Biochem. Biophys. Res. Commun. **182**, 1002-1007 (1992).

[13] Moore, R., J. F. T. Glasgow, M. A. Bingham, J. A. Dodge, R. J. Pollitt, S. E. Olpin, B. Middleton, K. Carpenter: Long-chain 3-hydroxyacyl-coenzyme A dehydrogenase deficiency – diagnosis, plasma carnitine fractions and management in a further patient. Eur. J. Pediatr. **152**, 433-436 (1993).

[14] Naylor, E. W., L. L. Mosovich, R. Guthrie, J. E. Evans, H. Tieckelmann: Intermittent non-ketotic dicarboxylic aciduria in two siblings with hypoglycemia: an apparent defect in β-oxidation of fatty acids. J. Inher. Metab. Dis. **3**, 19-24 (1980).

[15] Przyrembel, H., C. Jakobs, L. Ijlst, J. B. C. de Klerk, R. J. A. Wanders: Long-chain 3-hydroxyacyl-CoA dehydrogenase deficiency. J. Inher. Metab. Dis. 14 674-680(1991).

[16] Roe, C. R., D. S. Millington, D. L. Norwood, N. Kodo, H. Sprecher, B. S. Mohammed, H. Schulz, R. McVie: 2,4-Dienoyl-coenzyme A reductase deficiency: a possible new disorder of fatty acid oxidation. J. Clin. Invest. **85**,1703-1707 (1990).

[17] Schulz, H.: Beta-oxidation of fatty acids. Biochem. Biophys. Acta **1081**,109-120 (1991).

[18] Sewell, A. C., S. W. Bender, S. Wirth, H. Munterfering, L. Ijlst R. J. A. Wanders: Long-chain 3-hydroxyacyl-CoA dehydrogenase deficiency: a severe fatty acid oxidation disorder. Eur. J Pediatr. **153**, 745-750 (1994).

[19] Sorrentino, D., D. Stump, B. J. Potter, R. B. Robinson, R. White, C. L. Kiang, P. D. Berk: Oleate uptake by cardiac myocytes is carrier mediated and involves a 40 kD plasma membrane fatty acid binding protein similar to that in liver, adipose tissue and gut. J. Clin. Invest. **82**, 928-935 (1988).

[20] Spener, F., T. Börchers, M. Mukherjea: On the role of fatty acid binding proteins in fatty acid transport and metabolism. FEBS Letters **244**, 1-5 (1989).

[21] Stremmel, W.: Fatty acid uptake by isolated rat heart myocytes represents a carrier-mediated transport process. J. Clin. Invest. **81**, 844-852 (1988).

[22] Treem, W. R., C. A. Stanley, D. E. Hole, H. B. Leopold, J. S. Hyams: Hypoglycemia, hypotonia, and cardiomyopathy: the evolving clinical picture of long-chain acyl-CoA dehydrogenase deficiency. Pediatrics **87**, 328-333 (1991).

[23] Wanders, R. J. A., M. Duran, L. Ijlst J. P. de Jager, A. H. vyn Gennip, C. Jakobs, E. Dorland, F. J. vanSprang: Sudden infant death and long-chain 3-hydroxyacyl-CoA dehydrogenase. Lancet 11: 52-53 (1989).

[24] Wanders, R. J. A., L. Ijlst, F. Poggi, J. P. Bonnefont, A. Munnich, M. Brivet, D. Rabier, J. M. Saudubray: Human trifunctional protein deficiency: a new disorder of mitochondrial fatty acid β-oxidation. Biochem. Biophys. Res. Commun. **188**, 1139-1145 (1992).

[25] Wilcken, B., K. C. Leung, J. Hammond, R. Kamath, J. V. Leonard. Pregnancy and fetal longchain 3-hydroxyacyl coenzyme A dehydrogenase deficiency. Lancet **341**, 407-408 (1993).

[26] Wisneski, J. A., E. W. Gertz, R. A. Neese, M. Mayr: Myocardial metabolism of free fatty acids. J. Clin. Invest. **79**, 359-366 (1987).

5 Cardiomyopathy in β-ketothiolase deficiency

V. Hesse, A. C. Sewell, H. Böhles, H. Haberland, B. Middleton, B. Fiedler, H. Förster, W. Jänisch

β-ketothiolase deficiency (McKusick 20375) is an inborn enzyme deficiency of isoleucine and ketone body catabolism [24] presenting as an organic acidaemia [15, 16]. The disease was first described in 1971 [4, 5] and the enzyme defect proven by Robinson et al. in 1979 [24]. Inheritance is autosomal recessive.

Different enzymes are summarized under the designations of β-ketothiolase, 3-oxothiolase, 3-ketothiolase or β-ketoacyl-CoA-thiolase and acetoacetyl-CoA thiolase. There is a non potassium dependent ketoacyl-CoA-thiolase (EC 2.3.1.9.) specific for acetoacetyl-CoA localized in the cytoplasm of liver and brain as well as two mitochondrial thiolases (type I and type II). Type I thiolase (EC 2.3.1.16) shows a long-chain length specificity for 3-ketoacyl-CoA substrates. The mitochondrial type II thiolase (EC 2.3.1.9.) is potassium dependent and specific for the short-chain 3-ketoacyl-CoA substrates acetoacetyl-CoA and 2-methyl-acetoacetyl-CoA. The two mitochondrial thiolases are present in most tissues with the exception of brain where only a thiolase activated by potassium (EC 2.3.1.9.) exists in addition to the cytoplasmatic thiolase [17, 18, 19, 20, 28]. In liver tissue all three thiolases are present [6].

When β-ketothiolase is lacking there is no thiolytic cleavage of 2-methylacetoacetyl-CoA into acetyl-CoA and propionic acid (Fig. 1). As a consequence a disturbance of the intramitochondrial isoleucine and fatty acid degradation occurs resulting in a disturbance of energy formation as well as increased ketogenesis. Acetyl-CoA deficiency leads to substrate deficiency of the citric acid cycle. In addition the metabolism of long-chain fatty acid-CoA is disturbed at the level of the cleavage of acetoacetyl-CoA to acetyl-CoA (Fig. 1). Ketone body degradation through acetoacetyl-CoA is stopped and the equilibrium reversed leading to increased ketone body formation. This is further increased in cases of an enhanced metabolism of activated long-chain fatty acids, e.g. during infection. β-ketothiolase plays an important role in the oxidation and energy formation of ketone bodies as well as of fatty acids [24]. In this context acetyl-CoA formed from fatty acids, amino acids and pyruvate has a key position and we should remember that free fatty acids, glucose, lactate and "activated" ketone bodies are the most important substrates for cardiac energy metabolism. In patients with β-ketothiolase deficiency type 11, very low enzyme activities are found in fibroblasts and leukocytes [18]. The molecular defect in β-ketothiolase deficiency (EC 2.3.1.9) is very heterogeneous, which may explain the differences in both clinical presentation and biochemical pattern. This heterogeneity makes prenatal diagnosis at the DNA level very difficult [9]. The mitochondrial enzyme deficiency leads to an increased formation of different organic acids from isoleucine degradation namely 2-methylacetoacetic acid, 2-methyl-3-OH-butyric acid and tiglylglycine (conjugate of tiglic acid and glycine) as well as to an increase in acetic acid and 3-OH-butyric acid, resulting from the block of ketone body utilisation. The mutation mainly relates to the extrahepatic ketothiolase type II. Liver contains another isoenzyme for the acetoacetyl-CoA synthesis, which is not affected by the mutation [18].

Clinically, patients with β-ketothiolase deficiency present with severe intermittent ketoacidotic attacks mainly during infections, starvation or intake of protein rich food. Patients with a short-chain β-ketothiolase deficiency show symptom-free intervals because the long-chain thiolase to a certain extent is able to cleave acetoacetyl-CoA [24]. Children generally begin to present clinical symptoms between 6-20 months of age. However, earlier or later manifestation is possible. When breast fed, infants are generally without clinical symptoms, because the isoleucine concentration of breast milk is lower than that of cow's milk [7]. During a metabolic crisis a pathological excretion of the following "marker" substances is increased: 2-methyl-3-OH-butyric acid, 2-methylacetoacetic acid, tiglylglycine and n-butanone. Between ketotic crises urinary excretion may be normal or dependent on the protein intake. Unrecognized and untreated, this defect may lead to irreversible damage and even death.

The early diagnosis in cases of unclear ketosis is of crucial importance. However, up to now only some cases have been described and the variability of symptoms, originating from the localisation of this defect in different tissues, is not sufficiently well known. Equally no morphological alterations of the cerebrum have been reported until now.

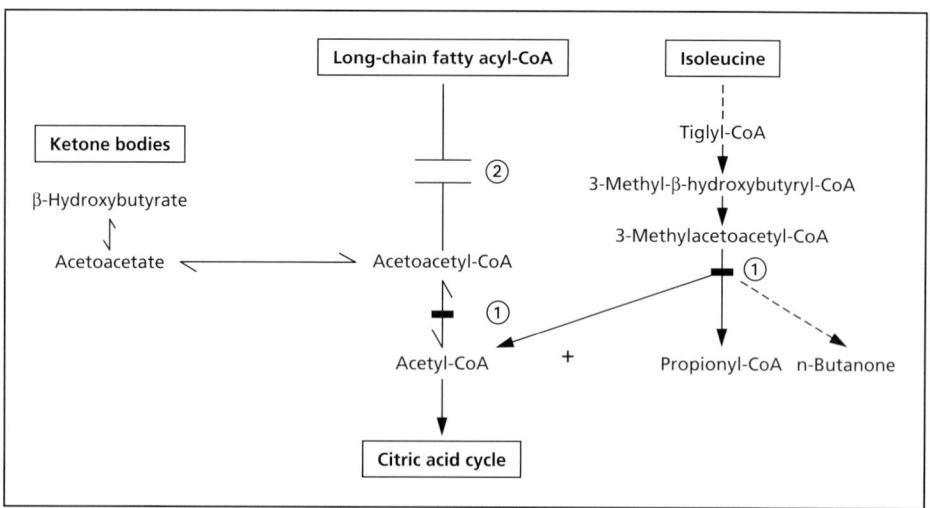

Fig. 1: Disturbances of ketone body and fatty acid oxidation. 1 = short-chain 3-ketoacyl-CoA thiolase; 2 = short- and long-chain 3-ketoacyl-CoA thiolase (β-oxidation of fatty acids).

Cardiomyopathy as a symptom was only described in one patient [11], but no electron microscopic analyses were reported.

5.1 Case report

We report an two brothers with β-ketothiolase deficiency who were diagnosed very late. They presented with a severe neurological disturbance and an obstructive cardiomyopathy. Both patients died in a severe metabolic crisis with respiratory infections.

5.1.1 Family history

The mother of the children was the second of 5 children and was reported to have developed normally until the age of 36 years, when neurological symptoms (walking difficulties, unsteady gait, blurred speech, unsecure finger-nose and heel-knee tests as well as disturbances of concentration and unstable affection) appeared. Multiple sclerosis was suspected and the patient was admitted for diagnostic work up. The presumptive diagnosis could not be verified either neurologically or by CSF analysis. CAT scan showed signs of a minor cortical atrophy. However in her urine traces of 2-methyl-3-OH-butyric acid and tiglylglycine were found. In skin fibroblasts a decreased β-ketothiolase activity into the heterozygote range was demonstrated (Tables 1 and 2). One sister of the patient's mother (4th of 5 children) is mentally handicapped. She is reported to have not walked before 3 years of age and is generally retarded in her psychomotor development, however she is able to live independently. The father has hypercholesterolaemia but otherwise is reported to be active and healthy.

5.1.2 Patients

The diagnosis of β-ketothiolase was made in the older brother (H. T.) (Fig. 2) at the age of 6 $^{9}/_{12}$ years following the analysis of the organic acid profile in the urine and in the younger brother (H. P.) at the age of 3 $^{9}/_{12}$ years following an isoleucine load (100 mg/kg) (Table 1). In the younger patient H. P. and his mother the diagnosis was verified by analysis of the enzyme activity (EC

Tab. 1: Urinary excretion of 2-methyl-3-hydroxybutyric acid and tiglylglycine in two brothers with β-ketothiolase deficiency and their parents.

Patient	Age	2-Methyl-3-hydroxybutyrate (mmol/mol creatinine)	Tiglylglycine (mmol/mol creatinine)
• H. T. basal	6.9	290	199
• H. P. basal	3.9	trace	n. d.
Post Ileu load*		480	1031
• H. C. (mother) basal	37	trace	trace
Post Ileu load*		trace	trace
• N. M. (father)	31	n. d.	n. d.
• Controls		<10	<2
* Oral loading with 100 mg/kg isoleucine		n. d. = none detected	

Tab. 2: **Ketolytic enzyme activities in fibroblast extracts** (nmol per mg protein; analyses B. Middleton, Univ. of Nottingham, UK).

Patient	Thiolase activity wich different 3-oxoacyl-CoA substrates (x ± SD)		Ratio of thiolase activities C_5/C_4 rate	Citrate synthase activity (x ± SD)
	2-Methylaceto-acetyl CoA (C_5)	Acetoacetyl CoA (C_4)		
• H. P. (4 years)	30.0	68.2	<u>0.44</u>	87.0
• H. C. (mother)	33.1	61.4	0.54	102.2
• Controls (n = 31)	49.9 ± 26.4	43.5 ± 21.1	1.18 ± 0.35	123.14- 51.6
• Heterozygotes for thiolase deficiency (n = 12)	8.0 ± 9.2	20.0 ± 14.9	0.34 ± 0.23	99.6 ± 42.4

2.3.1.9) in skin fibroblasts using 2-methylacetoacetyl-CoA (C5) and acetoacetyl-CoA (C4) as substrates. Mitochondrial citrate synthetase served as reference enzyme (Table 2).

Clinically both children presented with a regression of acquired abilities and showed an "extrapyramidal type" of tetraparesis, myoclonic-astatic seizures and amaurosis. Periods of metabolic acidosis occurred repeatedly, especially during infections of the upper respiratory tract.

The most representative parts of the history and results can be presented as follows: The patient was born at term. Weight: 3150 g; length: 50 cm; Apgar score: 9, 10, 10, 10. In the newborn period he presented with a meconium plug syndrome. Statomotor development was delayed. He could not sit before 12 months of age and walk before 2 years of age. At the age of $2^{10}/_{12}$ years the patient could not speak, however reacted to acoustic stimulation, presented with a grand-mal seizure and developed a crossed extrapyramidal form of tetraparesis. Myoclonic astatic seizures developed affecting the tongue and floor of the mouth as well as optic atrophy and retinitis pigmentosa. Retinal vessels could not be detected. Finally a state of bulbar paralysis was reached. The EEG showed signs of general disturbances including generalized spikes. Evoked visual potentials were pathological indicating damage of the pontine area. The acoustic nerves showed normal potentials. At the age of $2^{11}/_{12}$ years a CAT scan was performed which showed diffuse cortical and subcortical atrophy pronounced an the left side. Cardiomyopathy was first noticed at $4^{4}/_{12}$ years of age. A biopsy of the quadriceps femoris muscle showed muscle fibres of changing thickness without signs of myopathy or mitochondrial disease. Urinary organic acid analysis revealed pathologically elevated concentrations of 2-methyl-3-hydroxybutyric acid and tiglyl glycine (Table 1). Amino acids and oligosaccharides were normal. The patient was treated with a protein reduced diet (0.7-1.0 g/kg/day); carnitine (100 mg/kg/day) and anticonvulsive drugs (primidone; clonazepame) as well baclofen. The patient died during an acidotic metabolic crisis during an upper respiratory tract infection. At this time height was 117 cm (SDS: -1) and weight 16 kg. Patient H. P. born 12.6.1988; died 1.8. 1992.

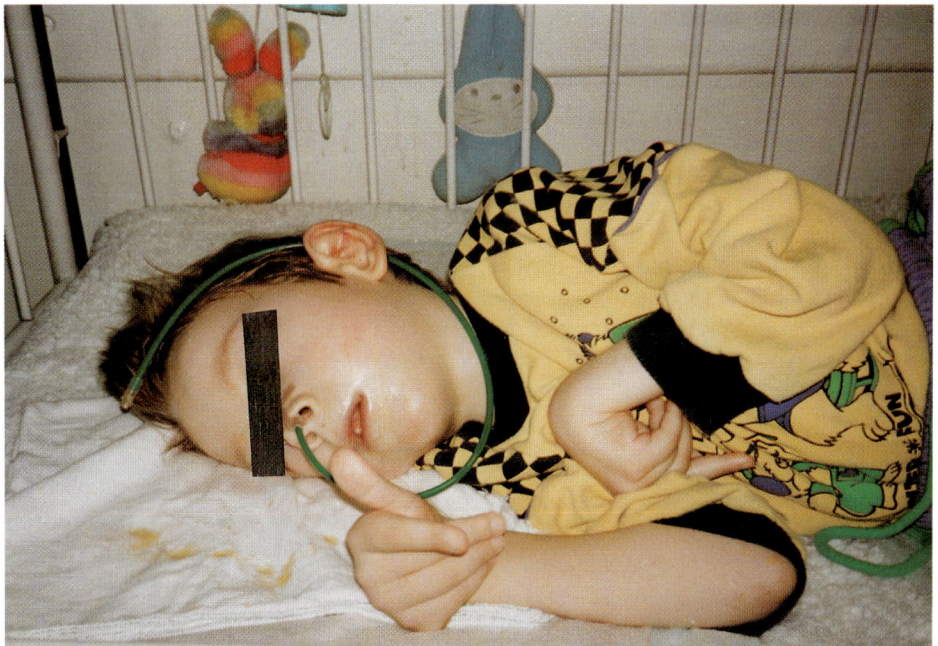

Fig. 2: Patient H. T. with untreated β-ketothiolase deficiency at age of 6 $^{9}/_{12}$ years. Amaurosis, tetraparesis, contractures, dysphagia, pseudobulbar paralysis.

The patient was born at term; weight 3270 g, length 50 cm. He was entirely breast fed for the first 6 months. The statomotor development during early life was normal. First tooth appeared at 6 months; sat free at 8 months and stood at 12 months. At the age of 18 months decreased motor abilities were first noticed. He was unable to stand, sit or turn over. Generalized hypotonia developed, followed at the age of 2 $^{7}/_{12}$ years by a myoclonic astatic seizure. He also had horizontal nystagmus, bilateral optic and retinal atrophy with fine granular pigmentation; persistent fisting, tetraparesis and myoclonus of the right arm; lacking head control and no reaction to acoustic stimuli. At 3 $^{9}/_{12}$ years the patient was amaurotic. At 4 $^{1}/_{12}$ years cardiomegaly was observed. At that time muscle reflexes of the lower extremities were symmetrically decreased and those of the upper extremities and abdomen could not be elicited. Urinary organic acid analysis showed only traces of 2-methyl-3-hydroxybutyric acid; however, following an isoleucine load high concentrations of 2-methyl-3-hydroxybutyric acid and tiglylglycine were formed (Table 1). The acid base status showed a metabolic acidosis: pH 7.25; HCO_3^- 14.7 mmol/l; base excess - 18 mmol/l.

The patient developed hepatomegaly and died in acidotic cardiac decompensation. At the time of death the patient was short, height 105 cm (-2.75 SD according to Hesse et al. [12]) and microcephalic.

Comparing the history of both brothers it is remarkable that essential symptoms had developed nearly at the same time. Seizures were first observed at $2^{10}/_{12}$ respectively $2^{7}/_{12}$ years of age. Amaurosis was diagnosed at $2^{10}/_{12}$ and $3^{8}/_{12}$ years of age and signs of retinitis pigmentosa were seen at $3^{8}/_{12}$ and $3^{9}/_{12}$ years of age. Cardiomegaly and cardiomyopathy were observed at $4^{4}/_{12}$ and $4^{1}/_{12}$ years of age. The boys died at $6^{10}/_{12}$ and $4^{2}/_{12}$ years of age respectively.

5.1.3 Neurological status

The neurological alterations were characterized by a selective disappearance of ganglia in the area of the caput nuclei caudati, putamen, claustrum and area 17 (visual field) as well as in the parasaggital zone of the parietal and occipital cortex. The chiasma opticum and the central visual tract showed demyelinisation [13]. MRI in the younger brother presented an increased signal intensity in the putamen area as sign of affected basal ganglia.

5.1.4 Cardiological status

Patient H. T.:

Cardiomegaly, extending to the left thoracic wall, with pulmonary congestion was first observed at the age of $4^{1}/_{2}$ years. The enlargement was related to all areas of the heart. The patient presented tachycardia and a $^{2}/_{6}$ systolic murmur above the 4-5 intercostal space in the left medioclavicular line. Blood pressure was within normal limits (Fig. 3).

The ECG showed a steep electrical axis with permanent sinus tachycardia of 120 bpm as well as signs of a beginning left and right heart congestion (R/S-ratio V1 = 33%, V6 = 62.5%). During digoxin therapy the cardiac enlargement increased and the ECG revealed an intensified cardiac hypertrophy. At autopsy the heart exhibited an extreme hypertrophy of the right and the left ventricle and also the septum. The aortic outflow tract was moderately obstructed. The total weight of the heart was significantly increased (285 g; age related normal 127 g).

Patient H. P.:

In this patient an extreme cardiomyopathy developed rapidly. The shape of the heart was still totally normal at the age of $3^{8}/_{12}$ years. The apex and the left ventricular wall, however, were slightly enlarged and interpreted as sign of a beginning hypertrophy. ECG showed a steep electrical axis; heart rate: 88 bpm. As the first sign of a cardiomyopathy a flattening of the T-wave in leads II, III, aVL and aVF was observed. As a sign of hypertrophy of the left ventricle and septum deep peaks of the Q-wave in leads V5 and V6 were recorded. An extreme cardiomegaly developed within 3 months. The heart diameter was enlarged. A $^{1}/_{6}$ systolic murmur could be heard in the 4-5 intercostal space. The ECG now presented overt signs of a ventricular hypertrophy including a left-sided bundle branch block.

At autopsy a distinct hypertrophic cardiomyopathy, particularly of the left heart and the septum including a slight obstruction of the aortic outflow tract, was demonstrated. The weight of the heart was increased to 125 g (age related normal 84 g).

Histology showed an increased amount of heart muscle cells with hyperchromatic nuclei in 30-50% of the cells (Fig 4). Deposits of granular PAS-positive material of hitherto unknown composition were present. In particular there were remarkable changes in heart muscle cell size (cellular hypo- or atrophy?) as well as lipid deposits in the perinuclear area (patient H. P.) and an increased interstitial cellularity in the left heart muscle (patient H. T.).

In the older brother a biopsy of the quadriceps femoral muscle was performed and no signs of myopathy were found.

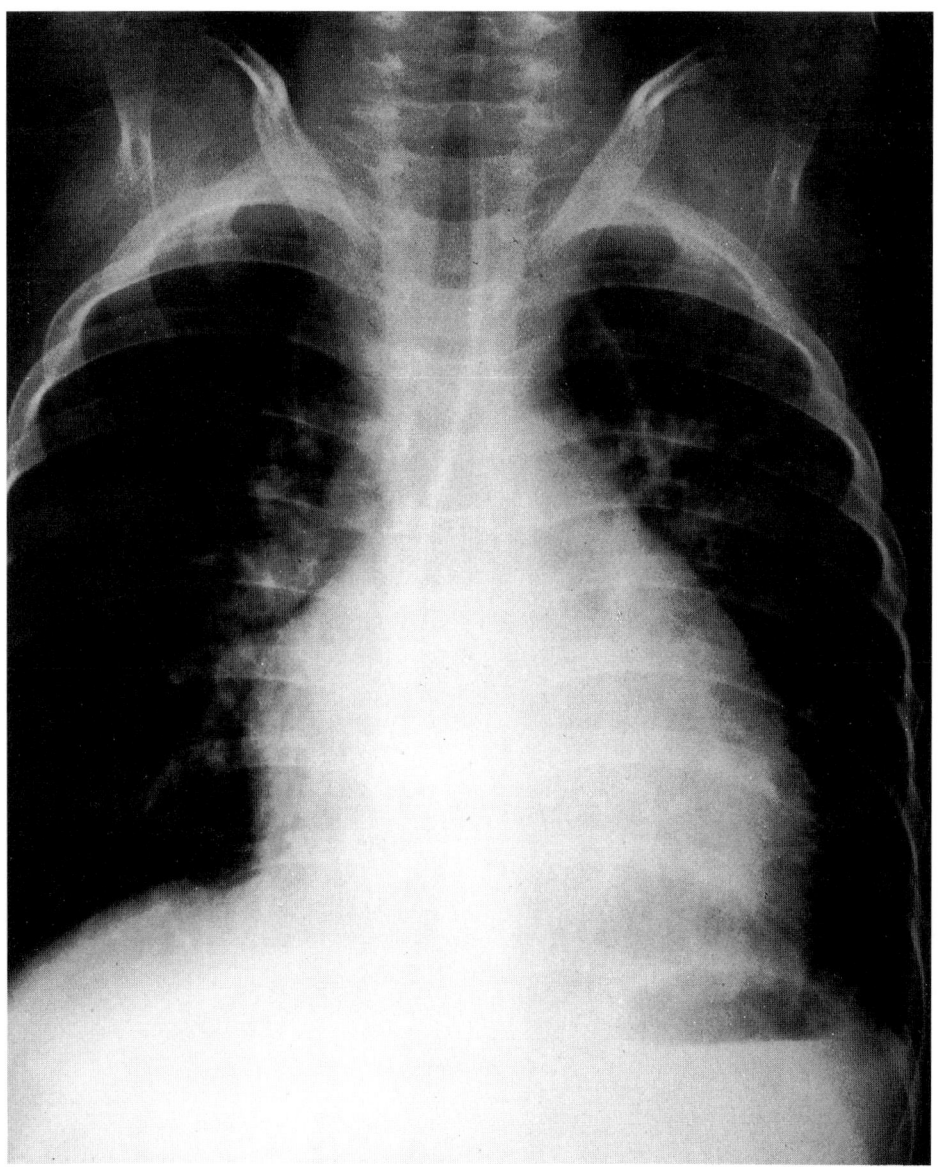

Fig. 3: Heart size in patient H. T. at the age of 4 $^{1}/_{12}$ years. Signs of cardiomyopathy and cardiac congestion.

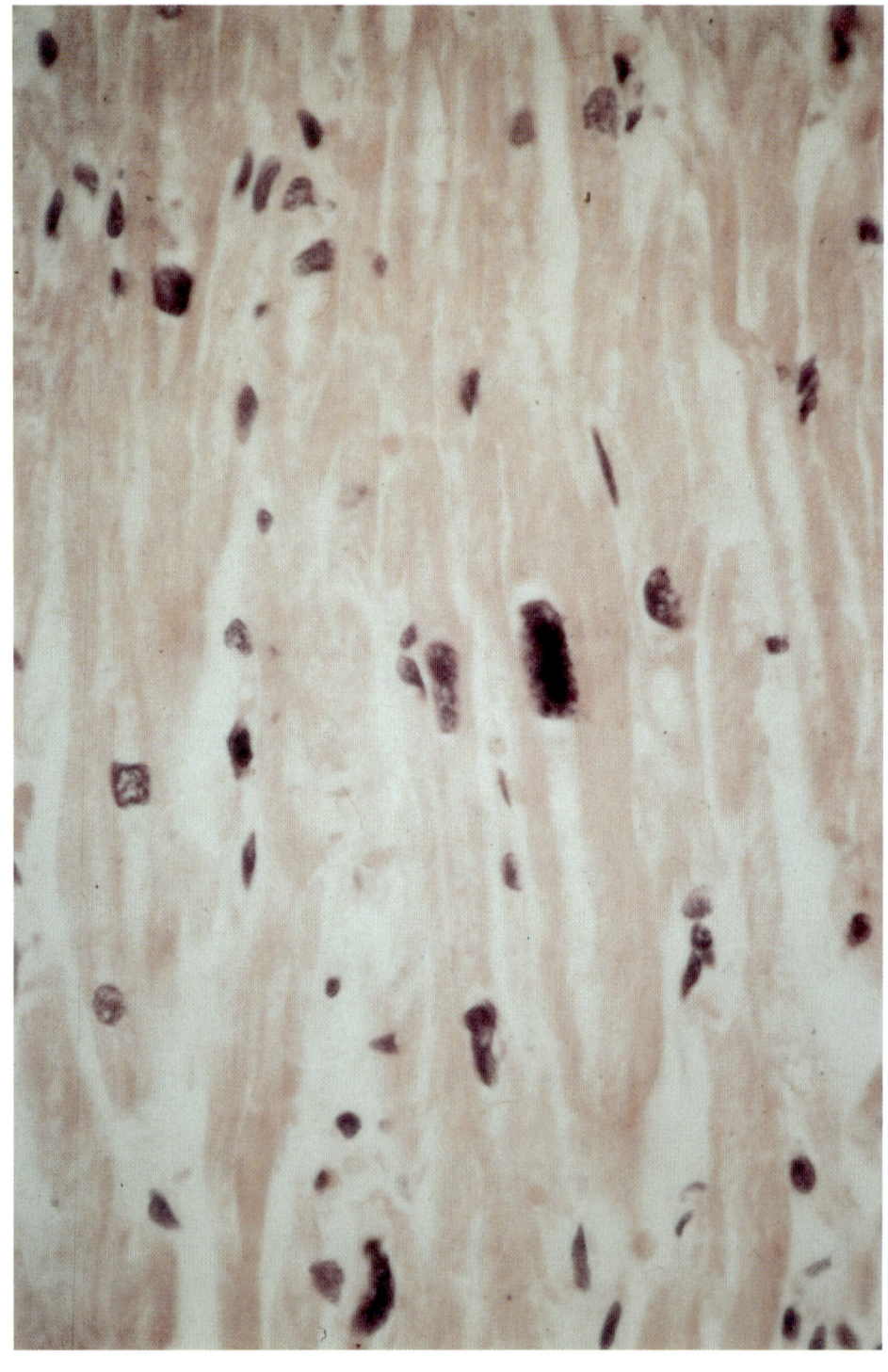

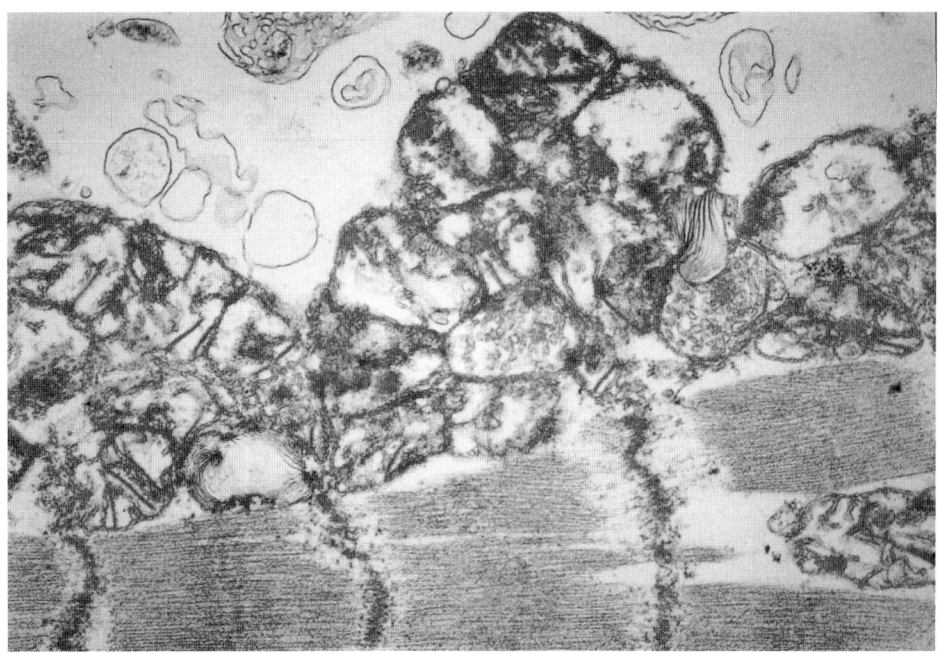

Fig. 5: **Multiple lysophagosomes.** Preserved Z-areas of the muscle fibres.

On electron microscopy, the presence of lysophagosomes (patient H. T.) indicated degradation of particles, possibly caused by toxic reactions or storage mechanisms (Fig. 5).

Cardiomyopathy may be part of different genetic and metabolic disorders [10]. Often the basis is a disturbance of mitochondrial metabolism [2, 8]. According to our present knowledge, congestive cardiomyopathy was hitherto described only once [11].

Apart from cardiomyopathies caused by mitochondrial disease like cytochrome-c-oxidase deficiency [14] or cytochrome-b-deficiency [23], they also may be due to myopathic carnitine deficiency [3, 27]. In patient H. P. and his parents the serum carnitine concentrations were normal as was the concentration of long-chain acylcarnitines in the heart muscle (Table 3). The patient was treated with carnitine on the assumption that accumulated toxic compounds, like tiglic acid, might be better excreted as their carnitine esters [22].

In the differential diagnosis long-chain 3-OH-acyl-CoA dehydrogenase deficiency was considered because it may cause cardiomyopathy; however, in both the patients no 3-OH-dicarboxylic acids were found in the urine. Urinary lactate excretion was also normal. The aetiology of the cardiomyopathy in β-ketothiolase deficiency is hitherto unknown, however it is suggested that it

Fig. 4: **Left ventricular cardiac tissue (patient H.T.).** Enlargement and nuclear hyperchromasia (about 30-50% of total nuclei) as a result of muscle fibre hypertrophy. Irregular fibre thickness. Perinuclear halo formation. PAS positive material in some areas.

Case report

Tab. 3: **Carnitine concentrations in serum and heart muscle in a boy with β-ketothiolase deficiency and obstructive cardiomyopathy.**

Serum	Pat. H. P.	Mother	Father	Controls
• Total carnitine (µmol/L)	44.8	36.6	44.8	53.8-77.5
• Free carnitine (FC) (µmol/L)	31.2	28.8	37.4	25.4-54.1
• Acyl carnitine (AC) (µmol/L)	13.6	7.8	7.4	5.4-30.1
• AC/FC	0.44	0.27	0.20	< 0.7
Heart muscle	**H. P. (right ventricle)**		**H. P. (left ventricle)**	
• Long-chain carnitine (nmoUmg NCP*)	2.06 (Normal = < 7.0)		1.39	

* NCP = non-collagenous protein

may result from a disturbed oxidative phosphorylation. In the model of the cardiomyopathic Syrian hamster, decreased activities of the enoyl-CoA hydratase, 3-hydroxyacyl-CoA dehydrogenase and β-ketothiolase, which are important for the intramitochondrial metabolism of fatty acids and isoleucine, were found in homogenates of heart muscle [1]. In the rat heart it was demonstrated that the myocardial β-ketothiolase activity is very high [20] and the oxidation of the branched-chain amino acids is of special importance in muscle metabolism [21].

The clinical symptoms of our patients, like psychomotor and mental retardation, seizures, ataxia, choreoathetosis, tetraparesis, blindness, cardiomegaly, atrophy of the optic nerve, anomalies of retinal pigmentation, cerebral atrophy and demyelinisation, were reminiscent of mitochondrial encephalomyopathies [25, 26]. The disturbance of mitochondrial energy production in β-ketothiolase deficiency may particularly effect energy dependent tissues, like heart muscle and basal ganglia.

References

[1] Barakat, H., W. Brown, S. D. Henry: Studies of fatty acid oxidation in homogenates of the cardiomyopathic hamster. Life Sci **23**, 1935 (1978).
[2] Böhles, H., H. Singer, W. Ruitenbeek, J. M. F. Trijbels, R. C. A. Sengers, U. P. Ketelsen, E. Wagner-Thiessen, H. Wick. Foamy myocardial transformation in a child with a disturbed respiratory chain. Europ. I. Pediatr. **146**, 582 (1987).
[3] Böhles, H. J.: Primäre und sekundäre Carnitinmangelzustände. In: K. Stehr, H. J. Böhles, Stoffwechselerkrankungen im Kindesalter, perimed Fachbuch Verlagsgesellschaft Erlangen, 224-231 (1987).
[4] Daum, R. S., P. H. Lamm, O. A. Mamer, C. R. Scriver A "new" disorder of isoleucine metabolism. Lancet ii, 1289 (1971).
[5] Daum, R. S., C. R. Scriver, O. A. Mamer, E. Delvin, P. Lamm, H. Goldman: An inherited disorder of isoleucine catabolism causing accumulation of a methylacetoacetate and a-methyl-β-hydroxybutyrate and intermittent metabolic acidosis. Pediatr. Res. **7**, 149 (1973).
[6] De Groot, C. J., Luit-de Haan, G., Hulstaert, E., Hommes, F. A.: A patient with severe neurologic symptoms and acetoacetyl-CoA thiolase deficiency. Ped. Res. **11**, 1112 (1977).
[7] Fomon, S. J.: Infant Nutrition (W. B. Saunders) Philadelphia 1967.
[8] Fujita, M., H. B. Neustein, P. R. Leeru: Transvasular endomyocordial biopsy in infants and small children: myocardial fingings in 10 cases of cardiomyopathy. Hum. Pathol. **10**, 15 (1979).
[9] Fukao, T., S. Yamaguchi, H. Nagasawa, M. Kano, T. Orii, Y. Fujiki, T. Osumi, T. Hashimoto: Molecular cloning of cDNA for human mitochondrial acetoacetyl-CoA thiolase and molecular analysis of 3-ketothiolase deficiency. J. Inher. metab. Dis. **13**, 757-760 (1990).
[10] Goldstein, J. L., M. S. Braun: Genetics and cardiovascular disease in Braunwald, E. ed: Heart disease, Philadelphia, 1980, W. B. Saunders 1683-1722

[11] Henry, C. G., A. W. Strauss, J. P. Keating, R. E. Hillman: Congestive cardiomyopathy associated with β-ketothiolase deficiency. J. Ped. **99**, 754 (1981).
[12] Hesse, V., U. Jaeger, K. Kromeyer, K. Zellner, L. Bernhardt, A. Hofmann: Aktualisierte Wachstumsdaten (Körperhöhle und Körpergewicht) 0-16jähriger Kinder (Jeaner Studien). Z. Klin. Med. **45**,1121 (1991).
[13] Jänisch, W., V. Hesse, B. Fiedler, H. Förster, H. Böhles. Pathomorphologische Befunde bei Ketothiolasemangel. Zentralbl. Pathol. **139**, 245 (1993).
[14] Ketelson, U. P.: Metabolische Myopathien unter besonderer Berücksichtigung morphologischer Befunde am Skelettmuskel. In: K. Steht, H. J. Böhles (Hrsg.) Stoffwechselerkrankungen im Kindesalter, Kinderheilkunde und Jugendmedizin, perimed Fachbuch, Verlagsgesellschaft mbH Erlangen, 1987, 203-223.
[15] Lehnert, W.: Relative Häufigkeitsverteilung von Organoazidurien bei der Suche nach angeborenen Stoffwechselerkrankungen. In: K. Stehr, H. J. Böhles (Hrsg.): Stoffwechselerkrankungen im Kindesalter, Kinderheilkunde und Jugendmedizin, perimed Fachbuch-Verlagsgesellschaft mbH, Erlangen 1987, 113.
[16] Lehnert, W., H. Niederhoff, W. Kinzer, W. Schaper: Diagnostik, Klinik und Therapie des alpha-methylacetoacetyl-CoA-thiolase-Mangels. Therapiewoche **29**, 8712 (1979).
[17] Middleton, B.: The Oxoacyl-Coenzyme A thiolase of animal tissue. Biochem. J. **132**, 717 (1973).
[18] Middleton, B.: Acidemia, 3-ketothiolase-deficiency. In: M. L. Buyse, Birth defects encyclopedia, Blackwell Sc. Publ. **14**, (1990).
[19] Middleton, B., R. G. Gray, M. J. Bennett: Two cases of ß-ketothiolase deficiency: a comparison. J. Inter. Metab. Dis. **7**, Suppl. **2**, 131 (1984).

[20] Middleton, B.: The acetoacetyl coenzyme A thiolases of rat brain and their relative activities during postnatal devolopment. Biochem. J. **132**, 731 (1973).
[21] Miller, L. L.: The role of liver and the non-hepato tissues in the regulation of free amino acid levels in the blood. In: J. T. Holden: Amino Acid Pools, American Elseviar Publishing Company, New York, p. 708 (1962).
[22] Millington, D. S., C. R. Roe, D. A. Maltby: Characterization of new diagnostic acylcarnitines in patients with β-ketothiolase deficiency and glutaric aciduria type I using mass spectrometry. Biomedical and Enviroment. Mass Spectrom **14**, 711 (1987).
[23] Papadimitriou, A., H. B. Neustein, S. Di Mauro, R. Stanton, N. Bresolin: Histocytoid cardiomyopathy of infancy: deficiency of reductible cytochrom b in heart mitochondria. Pediatr. Res. **18**, 1023 (1984).
[24] Robinson, B. H., W. G. Sherwood, J. Taylor, J. B. Balfe, O. A. Mamer: Acetoacetyl CoA thiolase deficiency: A cause of severe ketoacidosis in infancy simulating salicylism. J. Ped. **95**, 228 (1979).
[25] Sengers, R. C. A., A. M. Stadhouders, J. M. F. Trijbels: Mitochondrial myopathies Eur. J. Pediatr. **141**, 192 (1984).
[26] Siemes, H.: Mitochondriale Myopathien und Enzephalomyopathien. Mschr. Kinderheilkd. **133**, 798 (1985).
[27] Van Dyke, D. H., R. C. Gripps, W. Markesberry, DiMauro: Hereditary carnitine deficiency of muscle. Neurology **25**, 154 (1975).
[28] Williamson, D. H., M. W. Bates, M. A. Pase, H. A. Krebs: Activities of enzymes involved in acetoacetate utilization in adult mammalian tissues. Biochem. J. **121**, 41 (1971).

6 Cardiac involvement in glycogen storage diseases

R. Santer, K. Ullrich

An involvement of the heart is a relatively frequent finding in some types of glycogen storage diseases (GSDs). Here, we present the biochemical and genetic bases and the clinical variability of this group of diseases. Cardiac manifestations of the different disease entities are presented in more detail.

6.1 Biochemical bases

6.1.1 Glycogen storage diseases

Glycogen is a high-molecular-weight, highly branched, spherical structure composed of numerous glycosidically linked glucose molecules with either α-1,4-linkage or, at branching points, α-1,6-linkages. It is formed by the action of two enzymes, glycogen synthase and amylo-1,4-1,6-transglucosidase (branching enzyme) and serves as a glucose storage pool with a low osmotic effect. In liver, this glucose depot plays an important role in the maintenance of glucose homeostasis for the whole organism while in muscle it mainly serves as an energy depot for ATP synthesis of the individual cell.

Glycogen synthesis occurs exclusively in the cellular cytoplasm while two degradation pathways of different cellular localisation are operative (Fig.1):
1. in lysosomes, glucose is released by acid glucosidase (also termed acid maltase) cleaving both α-1,4- and α-1,6-bonds.
2. in the cytoplasm, glucose is released from glycogen by the synergistic action of the phosphorylase system and amylo-1,6-glucosidase leading to the formation of glucose-1-phosphate. Only in glucose-producing tissues like the liver is free glucose subsequently liberated by the action of the microsomal glucose-6-phophatase system and transported across the cellular membrane by facilitative or vesicle-associated transport (Figs. 1 a, b).

Both degradation pathways are not functionally connected. Thus, a decreased activity of one of the enzymes involved leads either to lysosomal or cytosolic accumulation of glycogen. Normally, the glycogen content of liver and muscle is not greater than 7 % and 2 % of the wet tissue weight, respectively [1].

Functionally, GSDs with cytosolic accumulation of glycogen can be distinguished from a disturbed lysosomal degradation on the basis of their tendency to hypoglycaemia as the lysosomal

48 Biochemical bases

Type	Biochemical basis [tissue expression]*	OMIM Links -disease	OMIM Links -protein	Gene	Affected organs	Frequency **	
0	– **glycogen synthase** [liver]	240600	138570	GYS2	L	rare	
I	**glucose-6-phosphatase system** [liver, kidney]					25 %	
a	– glucose-6-phosphatase	232200	232200	G6PC	L,K		~85 %
non-a	– glucose-6-phosphate translocase	232220	602671	G6PT1	L,K		~15 %
II	**lysosomal glycogen storage disease** [generalized]					15 %	
a	– glucosidase	232300	606800	GAA	**H,M**		
b	– lysosome-associated membrane protein-2	300257	309060	LAMP2 ***	H,M,B		rare
III	– **glycogen "debranching enzyme"**	232400	232400			25 %	
a	[liver/muscle]			AGL	L,M,H		~85 %
b	[liver only]			AGL	L		~15 %
c	1,6-glucosidase activity only			AGL	L,M,H		rare
d	oligo-1,4-1,4-glucan transferase activity only			AGL	L,M,H		rare
IV	– **glycogen "branching enzyme"**	232500	607839	GBE1	L	3 %	
V	– **phosphorylase** [muscle]	232600	232600	PYGM	M	2 %	
VI	– **phosphorylase** [liver]	232700	232700	PYGL	L	rare	
VII	– **phosphofructokinase** [muscle]	232800	171860	PFKM	M, (H)	rare	
VIII	original designation for subgroups of type IX						
IX	**phosphorylase kinase**					25 %	
a-1	– subunit α_2 [liver, blood cells]	306000	306000	PHKA2 ***	L		~75 %
a-2	– subunit α_2 [liver]			PHKA2 ***	L		rare
	– subunit β [liver, blood cells, muscle]	261750	172490	PHKB	L,M		rare
c	– subunit γ_2 [testis, liver]		172471	PHKG2	L		rare
d	– subunit α_1 [muscle]		311870	PHKA1 ***	M		rare
cardiac	unknown	261740		?	**H**		rare
X	not uniformly used designation						
XI (FBS)	– **glucose transporter 2**	227810	138160	GLUT2	L,K	rare	
	+ **AMP activated protein kinase** subunit γ_2	600858	602743	PRKAG2	**H**	rare	

pathway does not contribute to glucose homeostasis. The degree of hypoglycaemia is different in disorders with an impaired cytosolic glycogen metabolism; it is most severe in defects of the glucose-6-phosphatase system in which both glucogenolysis and gluconeogenesis are impaired. Patients with defects of amylo-1,6-glucosidase or the phosphorylase system generally do not tend to present with severe hypoglycaemia because the cytosolic glycogen degradation is only partially disturbed and gluconeogenesis remains intact.

Some of the enzymes involved in glycogen degradation show an individual organ expression which explains why some GSD subtypes are restricted to specific tissues while others show an involvement of more than one organ (Table 1). As an example, the glucose-6-phosphatase system which is affected in GSD I is not expressed in muscle and heart tissue which explains why cardiomyopathy is not a symptom in GSD type I patients.

The glycogen debranching enzyme, the enzyme deficient in GSD III, has both oligoglycan-1,4-1,4-transferase and amylo-1,6-glucosidase (AGL) activity (Fig. 1b). It is expressed in different tissues and, in addition to liver and muscle, it can also be found in heart tissue. To the best of today's knowledge, there is only one gene coding for this enzyme from which both the liver and muscle debranching enzyme are transcribed [2]. Interestingly, however, mutations of exon 4 to 35 of this gene affect both liver and muscle (GSD IIIa) while mutations of exon 3 only affect the liver (GSD IIIb) [3]. This suggests that tissue specific differences of transcription exist [4].

The phosphorylases that contribute to glycogen degradation in muscle and in liver are encoded by two different genes [5,6]. Defects within these genes, GSD V and GSD VI, respectively, may lead only to the involvement of one organ. Heart involvement in muscle phosphorylase (GSD VI) has hitherto not been described.

Phosphorylases are strongly regulated by the activating enzyme phosphorylase kinase. This enzyme is a tetramer that exists in several tissue-specific isoforms. The β and the δ subunits (calmodulin) are identical in liver and muscle while tissue-specific subunits exist for the α and γ chain. Defects of different subunits of phosphorylase kinase have been described (Table 1) and these disorders represent the heterogenous group of GSD type IX. Within this group, patients with a combined involvement of muscle and heart have not been described. However, in a few patients an isolated cardiac phosphorylase kinase deficiency has been detected [7,8,9,10] which suggests that a specific, not yet identified subunit

---

Tab. 1: Systematic classification, frequency and cardiac involvement in different types of glycogen storage diseases. Grey shading highlights disease entities in which cardiomyopathy has been described.

*   L = liver, K = kidney, H = heart, M = muscle, B = brain
**   the first column gives the frequency among all types of glycogen storage diseases estimated on the basis of data from Hers and Fernandes, Groningen, NL, EU, and Chen, Durham, NC, USA; the second column shows the frequency of a GSD subgroup within the respective groups.
*** X-chromosomal inheritance, all other types show autosomal recessive inheritance

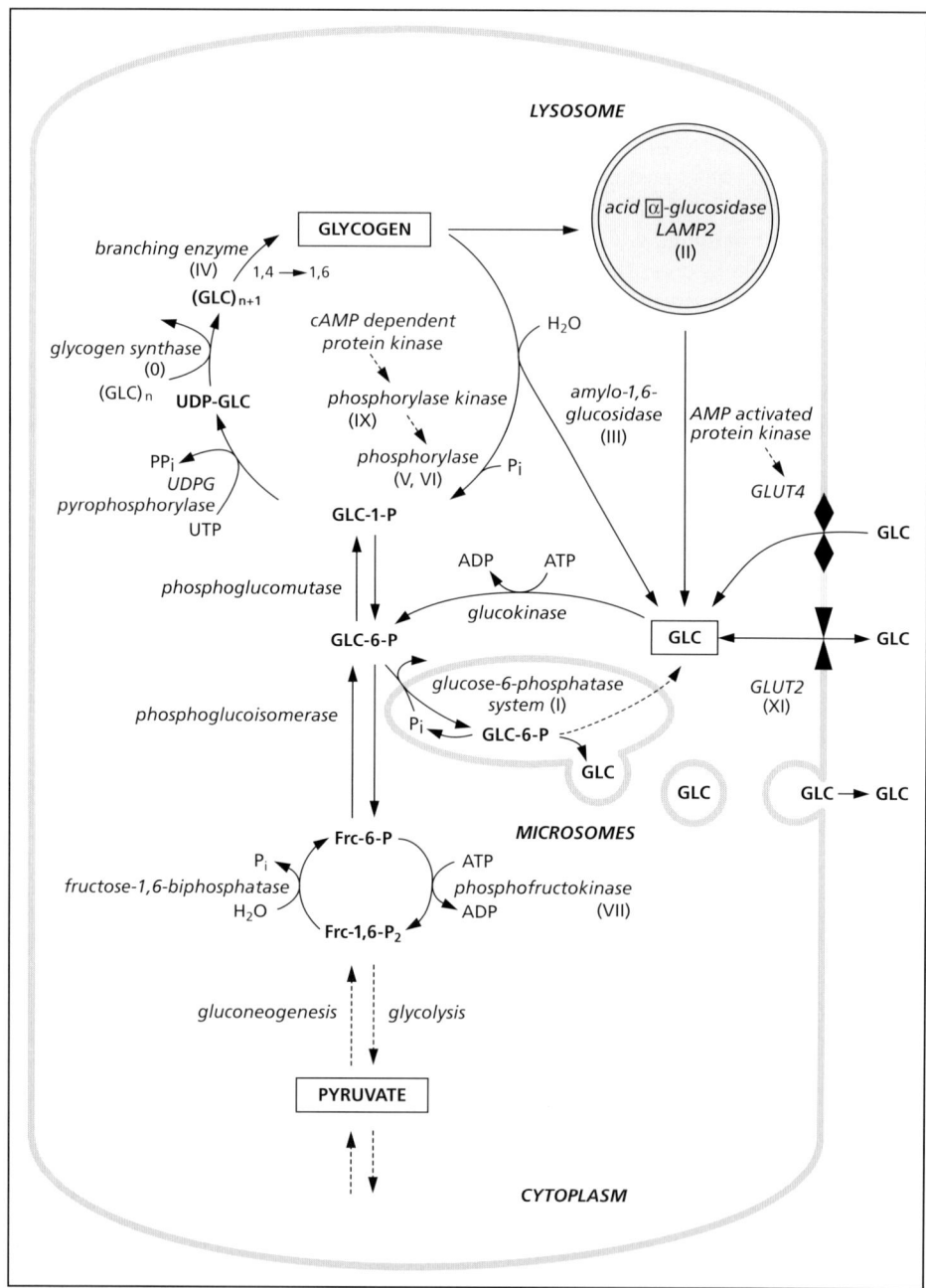

Fig. 1: a) Principal pathways in glucose and glycogen metabolism. Note that pathways, enzymes and transporters are differently expressed in different tissues (see biochemical basis). As an example, the microsomal glucose-6-phosphatase system responsible for the generation of free glucose is present in liver and renal tubular cells but not in muscle. The mayor glucose carrier in hepatocytes is GLUT2 while it is GLUT4 in muscle. Specific types of glycogen storage disease are shown in blue.

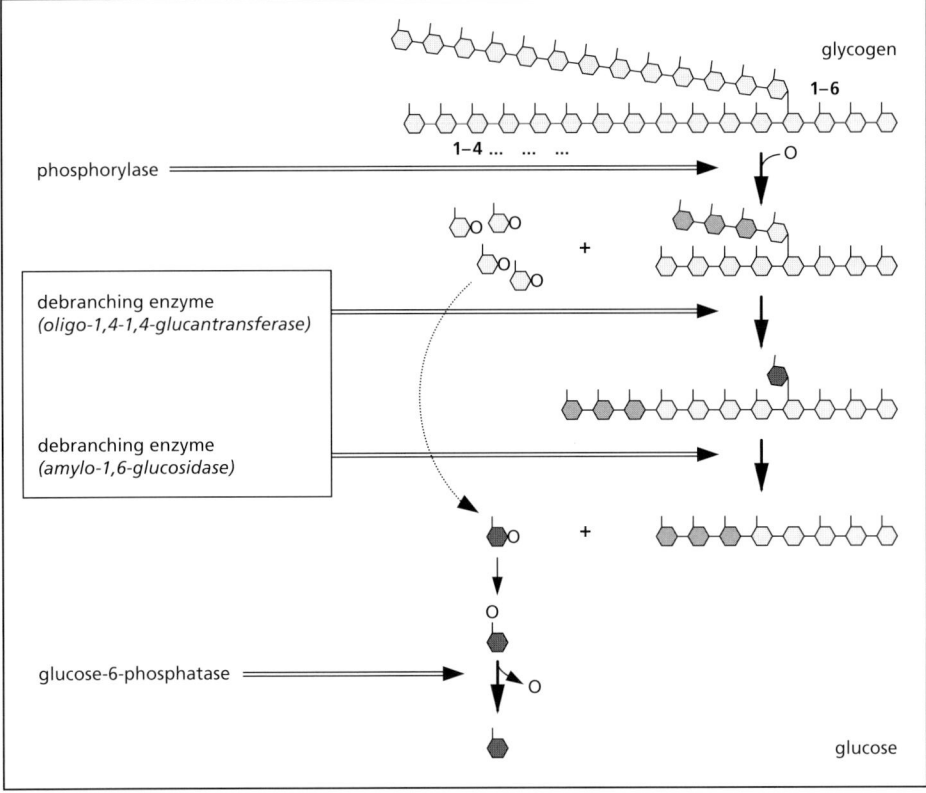

Fig. 1: b) Enzymatic steps of glycogen degradation.

must exist in heart tissue. Interestingly, the α-subunit of the myocardial phosphorylase kinase has been reported to have a lower molecular weight than that of skeletal muscle; however, it is still unclear whether both protein chains are encoded by different genes [11,12].

6.1.2 Disorders with vacuolar glycogen storage

As lysosomes are part of all cell species except mature erythrocytes, it is well conceivable that a decreased activity of lysosomal α-glucosidase, the enzyme affected in GSD IIa, is always associated with generalized storage phenomena in several tissues, i.e., both the heart and skeletal muscle. Tissue specific isoenzymes have hitherto not been described. Lysosomal enzymes are glycoproteins. On the way from their place of synthesis (the endoplasmic reticulum) to the lysosomes, the peptide structure and/or the carbohydrate side-chain are enzymatically modified or they undergo different steps of modification or maturation in prelysosomal and lysosomal cell compartments.

As a result of these modification steps, the mature intralysosomal enzyme has a lower molecular weight than the precursor enzyme. The exemplary modification steps of α-glucosidase are presented in Fig 2. In the Golgi complex, the carbohydrate side-chains are modified and phosphorylated to mannose-6-phosphates. This enables the protein to bind to mannose-6-phosphate-specific receptors and regulates the intracellular separation of lysosomal enzymes from secretory glycoproteins. The enzyme transport from the Golgi complex to the lysosomes occurs in vesicles that bind to two different mannose-6-phosphate-specific receptors of the lysosomal membrane. This membrane plays an important role in the function of lysosomes by sequestering numerous acid hydrolases that are responsible for the degradation of foreign materials and for specialized autolytic functions. Two glycoproteins, LAMP1 and LAMP2 constitute a significant fraction of the total lysosomal membrane glycoproteins. LAMP2, the protein deficient in GSD IIb, is a recognition molecule for autophagosomes which mediates the docking to and the fusion with lysosomes. If this protein is missing, glycogen will accumulate within autophagosomes because it cannot be degraded by lysosomal α-glucosidase [13,14,15].

The individual steps of modification within the cell are not identical for different lysosomal enzymes. In GSD IIa, discussed in this chapter, several mechanisms of mutations have been described that result in
- defective synthesis of the enzyme protein,
- normal or decreased synthesis of an unstable enzyme protein undergoing degradation during transport to the lysosomes,
- decreased enzyme phosphorylation, and
- synthesis of an enzyme protein with impaired catalytic function [16, 17, 18, 19].

Individual mutations result in a varying degree of functional impairment of the enzyme. Therefore, cultivated fibroblasts of patients with the classical infantile-type GSD IIa show an enzyme activity of less than 3% of controls and patients with the late-onset muscular type generally have a residual activity of 5-20%. Different mutations of the α-glucosidase gene in the homozygous state or in combination with others may occur. At least in part, this may explain the large variability of the individual clinical picture [20]. The different organ involvement in lysosomal disorders may be explained on the basis of several parameters such as residual activity in different tissues, different degrees of substrate accumulation as a consequence of synthesis and endocytosis, and functional disturbances caused by the storage material.

6.2 Clinical aspects

6.2.1 Disorders with cytosolic glycogen storage

6.2.1.1 Amylo-1,6-glucosidase deficiency, GSD III (Cori disease)

Most patients with GSD III present with a decreased debranching enzyme activity both in liver and muscle tissue (termed GSD IIIa, see Table 1), however, clinically liver involvement with hepatomegaly dominates in children and adolescents. Therefore, it is often difficult to distinguish these patients from those with GSD I, i.e. patients with defects of the glucose-6-phosphatase system. However, in GSD III hypoglycaemia is less frequent than in type I and it is not

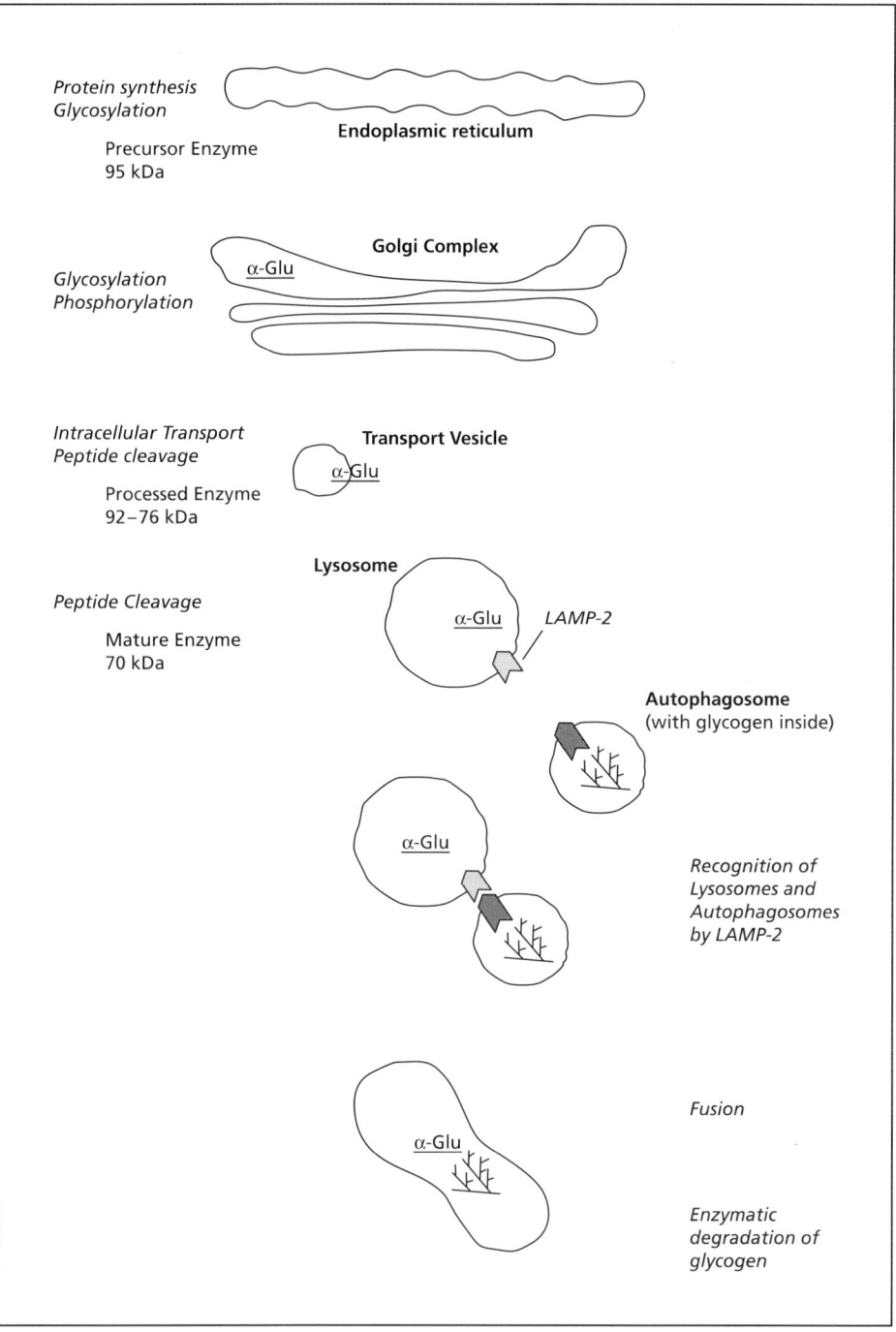

Fig. 2: **Intracellular transport and processing of acid α-glucosidase (α-Glu).** The role of LAMP-2 as a docking protein for autophagosomes is schematically presented (modified from [16]).

Tab. 2: Neuromuscular findings in GSD III

Symptoms / Tests	
▪ childhood onset	32 %
▪ muscle weakness	91 %
▪ muscle wasting	76 %
▪ distal > proximal	50 %
▪ pathological EMG	82 %
mixed features	53 %
▪ pathological nerve conduction velocity	46 %
▪ pathological ECG	82 %
▪ pathological forearm ischemia test	54 %

* from reference [24]

accompanied by an elevation of serum lactate. The kidneys are of normal size. Glycogen content in liver is elevated; in muscle it may be normal despite a demonstrable enzyme deficiency [21]. The definitive diagnosis is based on enzymatic studies in erythrocytes, fibroblasts, muscle, or liver tissue. Primary molecular diagnosis is possible but it is time consuming due to the fact that the gene coding for the debranching enzyme is relatively large and, with the exception of few small populations, common mutations have not been observed.

The clinical course of myopathy does not differ whether the entire enzyme protein is missing or whether only the glucosidase or transferase activity of the enzyme is reduced [22]. Early clinical signs of muscular involvement are statomotor retardation with late walking age, muscular hypotonia, unstable gait as well as difficulties when climbing stairs. Frequently, these symptoms are misinterpreted as the consequence of a central nervous problem. With beginning puberty, the severity of hepatopathy decreases whereas the involvement of skeletal muscle becomes increasingly prominent with advancing age [21,23]. Characteristics of neuromuscular symptoms of GSD III patients are listed in Table 2 [24]. As a consequence of an increased gluconeogenesis from protein, the muscular system can present as only weakly developed even in the presence of normal activity of amylo-1,6-glucosidase activity in muscle. Cases with primary muscle involvement are almost always associated with an increase in serum creatine kinase activity. However, this increase is not correlated with the severity of myopathy [22,25]. The increased activities of serum alanine aminotransferase and aspartate aminotransferase are predominantly caused by the associated hepatopathy [26].

Tab. 3: Frequency of cardiac involvement in GSD III *

Author	n	Patients' age [years]	%
Cornelio et al. 1984 [24]	22	?	82
Moses et al. 1989 [27]	20	3 – 30	95*
Smit et al. 1990 [28]	41	?	65
Labrune et al. 1991 [25]	18	2 – 30	33
Coleman et al. 1992 [22]	9	4 – 55	66
Carvalho et al. 1993 [29]	23	?	65
Talente et al. 1994 [30]	9	- 18	44

* see Table 4 for details

Systematic prospective investigations on the long-term course of cardiac changes have not been reported until now and only small series of personal observations exist [22, 24, 25, 27, 28, 29, 30]. Heart muscle involvement in GSD III was observed in 33 - 90 % of different series of clinical investigations (Table 3). Only rarely, the diagnosis was not established from the clinical course but from findings obtained during a cardiac biopsy [31]. Besides the myocardium, skeletal muscle has almost always been affected in cases of cardiomyopathy; although there has been no correlation between the extent of cardiac involvement and the severity of the skeletal muscle problems [22,25,27,32]. The cardiac involvement in GSD III manifests in the majority of cases in left or right ventricle and/or septum hypertrophy which can appear very severe upon clinical imaging. Functional studies, however, have shown that it is important to consider the aetiology of left ventricular hypertrophy before the application of risk stratification [33]. In GSD III patients, left ventricular hypertrophy has generally been considered to be of minor clinical significance and outflow obstruction is a rare problem [22, 25, 27, 30, 32, 33]. Myocardial fibrosis resulting in rigidity of the ventricle seems to be rare [34].

Tab. 4: Cardiac involvement in GSD III *

▪ clinical	1 / 20
▪ X-ray, cardiomegaly	2 / 20
▪ echocardiography	
- left ventricular hypertrophy	13 / 16
- left ventricular dilatation	2 / 16
- left atrial dilatation	6 / 16
▪ eletrocardiography	
- left-/biventricular hypertrophy	19 / 20

* from reference [27]

Table 4 summarizes the spectrum of cardiac manifestations observed in GSD III patients 3 to 30 years of age [27]. We have seen single younger patients with cardiac problems; some of them, however, were most likely not caused by the underlying biochemical defect but were precipitated by an inappropriate treatment with very high oligosaccharide or cornstarch application with the consequence of hyperinsulinaemia [35]. In five patients with an average of 9.5 years, Labrune et al. found no alterations of the echocardiographic parameters during a follow-up period of at least 3 years [25]. In addition, 3 adults have been reported with primary myopathy developing signs of heart insufficiency [21, 25, 27]. It has been reported that muscle force and ECG changes improve on a protein-rich diet [36] which we cannot confirm from our experience with 3 patients followed over several years.

6.2.1.2 Branching enzyme deficiency, GSD IV (Andersen disease)

GSD IV is caused by a genetic defect of the glycogen branching enzyme amylo-1,4-1,6-transglucosidase which is responsible for the formation of 1,6-linkages within the glycogen molecule. This disease is characterized by the deposition of an atypical glycogen molecule ('amylopectinosis') which shows a pathological iodine staining and which is electron microscopically defined by decreased ramification and a granular-fibrillar structure [21, 37, 38]. In these patients a progressive hepatosplenomegaly with development of liver cirrhosis and ensuing portal hypertension as well as hypotonia and atrophy of skeletal muscle are clinically prominent. Generally, patients with the early-onset type die within the first 4 years of life. The late-onset form, also termed 'adult polyglucosan body disease', is dominated by neuromuscular symptoms caused by deposits that can be detected in biopsies of the skin and skeletal muscle as well as in the peripheral and central nervous system. Only minor liver involvement has been described [21, 38, 39, 40]. Adult-type GSD IV usually has its manifestation at an age of 50 to 60 years. Branching enzyme activity might not be the only factor that determines the clinical course because also in these late-onset cases a complete absence of enzyme activity in several tissues can be found [40].

Cardiac storage of the atypical glycogen has repeatedly been demonstrated in postmortem studies and also in vivo by heart muscle biopsy; symptomatic heart disease, however, has only rarely been reported both for paediatric and adult cases [40, 41, 42, 43, 44]. The few GSD IV cases with cardiac involvement may present either with hypertrophic cardiomyopathy [41] or with dilated ventricles [40, 45, 46]. Single observations show that cardiomyopathy can be the very first symptom of generalized branching enzyme deficiency and that first symptoms can already occur in the neonatal period [46] but also beyond the age of 18 years [45].

Orthotopic liver transplantation has been proposed to treat patients with GSD IV because of early progressive cirrhosis. Few cases have been reported with absence of disease progression in other organs after liver transplantation and even regression of cardiac amylopectin infiltration [47,48]. Other observations contrast with this experience and have seen progressive cardiac glycogen deposition even after liver transplantation [49]. Heart transplantation has only been reported in one 17-year-old patient with cardiomyopathy secondary to GSD IV. After a 1-year follow-up, he was in good clinical condition with no visible recurrence of glycogen deposition in the donor organ on endomyocardial biopsy [50].

6.2.1.3 Phosphofructokinase deficiency, GSD VII (Tarui disease)

Muscle cramps with exertion and myoglobinuria upon extreme exertion are typical signs of GSD VII which is caused by a defect of the muscle (M) isoenzyme of phosphofructokinase while liver (L) and platelet (P) isoenzymes, which are under different genetic control, are not affected. GSD VII is usually associated with mild haemolytic anaemia since red cells normally contain phosphofructokinase tetramers which are normally composed of L and M monomers.

Cardiomyopathy has only been described in a single case of phosphofructokinase deficiency presenting with an unusual multisystem disease [51]. In this case, excessive cytosolic glycogen storage was found in heart and liver. Phosphofructokinase activity was lacking in both liver and muscle, and it has been suggested that this unexpected presentation may be related to the absence of an unknown activator common to all phosphofructokinase isoenzymes.

6.2.1.4 Phosphorylase kinase deficiency, GSD IX

Phosphorylase kinase-deficient glycogenosis of the liver manifests in infancy with hepatomegaly, growth retardation, and elevated plasma aminotransferases and lipids. It is most frequently caused by mutations in the gene coding for the α-subunit of phosphorylase kinase. It is usually a benign condition, often with complete resolution of symptoms during puberty. A minority of patients displays a more severe phenotype with symptomatic fasting hypoglycaemia and abnormal liver histology that may progress to cirrhosis. These patients with liver cirrhosis in childhood were repeatedly found to have mutations of the γ-subunit [52]. Phosphorylase kinase deficient patients with hepatic and muscular involvement had mutations of the β-subunit [53]. In none of these GSD IX subtypes has cardiac involvement been described. Conversely, a few patients with phosphorylase kinase deficiency confined to the heart have appeared in the literature. All three children reported died at an early age of 2-5 months as a consequence of their heart disease [7, 8, 9, 10]. The ECG showed a shortening of the PQ interval and in one patient signs of a pre-excitation syndrome. The electro- and echocardiographic signs of a biventricular hypertrophy were present. The pathoanatomical analysis revealed severe storage of normally structured glycogen [10] but also a biventricular endocardial fibroelastosis in one case. The 'lacework appearance of the myocardium' visible on light microscopy is similar to that generally found in GSD II. In contrast to GSD II, however, GSD IX patients with cardiomyopathy do not present with a general muscular hypotonia. The serum creatine kinase concentration was only slightly above normal in one patient.

6.2.1.5 Constitutive activation of AMP-activated protein kinase

Constitutive activation of AMP-activated protein kinase (AMPK) is a novel cause of a myocardial glycogen storage that has only recently been reported [54, 55]. Due to its ability to modulate GLUT4 translocation to the cellular membrane, this enzyme plays a pivotal role in the regulation of glucose uptake; AMP-activated protein kinase further regulates hexokinase activity and enzymes of glycolysis. AMPK is thus able to switch off ATP-consuming biosynthetic pathways. Because AMPK provides a central sensing mechanism that protects cells from exhaustion of ATP supplies, it has been proposed that energy compromise may provide a unifying pathogenic mechanism in all forms of hypertrophic cardiomyopathy [56].

AMPK is a heterotrimeric protein composed of the catalytic a subunit and the non-catalytic subunits β and γ. Mutations of the γ2 subunit (PRKAG2) expressed in heart muscle cells were originally detected in families with dominant types of ventricular pre-excitation syndromes [56, 57]. In addition to the propensity to arrhythmia, a significant proportion of these patients was reported to have hypertrophic cardiomyopathy [56]. Histological and biochemical investigations showed that these patients had myocyte enlargement due to vacuoles containing glycogen derivatives and only minimal interstitial fibrosis but that there was no myocyte or myofibrillar disarray [54].

Clinically, these patients usually presented early in life with tachyarrhythmia (atrial fibrillation, supraventricular arrythmia) sometimes associated with syncopes. Electrophysiological studies revealed accessory atrioventricular pathways. How glycogen accumulation or the underlying genetic defect accounts for the presence of these extra bundles has remained obscure. With increasing age, patients developed bradycardia and heart blocks requiring pacemaker implantation. Left ventricular hypertrophy was

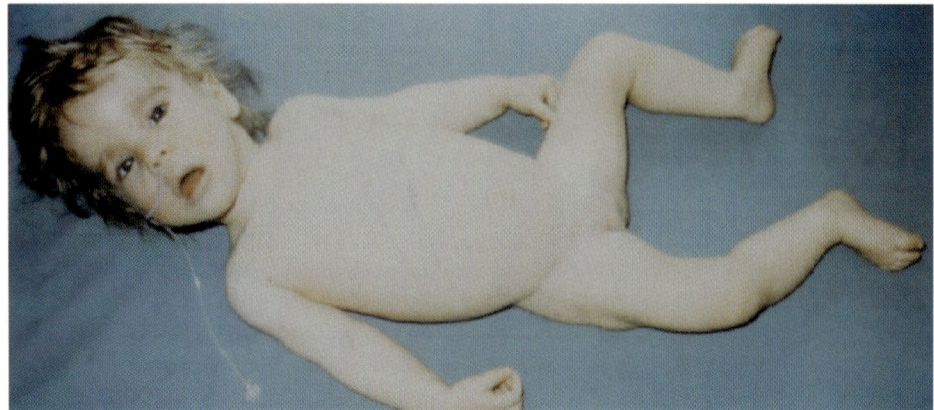

Fig. 3: Clinical aspect of an infant with infantile α-glucosidase defiency (Pompe's disease). Note the low muscle tone with an open mouth and the frog-like position of the legs. Nasogastric feeding is necessary because of poor sucking. Mild hepotomegaly is indicated.

present in more than 70% of patients in their thirties accompanied by significant deterioration of ventricular function. In a series of 70 patients, five required cardiac transplantation; in 4 out of 70 sudden death occurred [54].

6.2.2 Disorders with vacuolar glycogen storage

6.2.2.1 Lysosomal α-glucosidase deficiency, GSD II a

Lysosomal α-glucosidase deficiency, generalized type **(Pompe's disease)**

The clinical presentation of this type of lysosomal α-glucosidase deficiency is strikingly consistent. First symptoms like weak cry, lack of emotional expression, lack of movements, poor sucking, muscular hypotonia, dyspnoea, a greyish skin colour and other signs of respiratory and cardiac insufficiency are generally observed between the second and fourth month of life (Fig. 3). Although rare, individual patients may already be symptomatic directly after birth [58]. At the time of clinical diagnosis, almost all cases have cardiomegaly, and this is the leading sign in most cases. Only very rarely have exceptional cases been observed that have not developed cardiomegaly during the course of their disease [59, 60]. In approximately one third of the cases, a non-characteristic heart murmur is present. Only about 50% of cases present with minor hepatomegaly accompanied by abnormal liver function tests (unless this is not the result of heart failure), and about 30% of cases have macroglossia. There is no typical order of symptoms: cardiomegaly and respiratory insufficiency may develop before generalized muscular hypotonia and vice versa. Lysosomal storage further results in an involvement of the central nervous system, and as a consequence, a decrease of deep tendon reflexes, dysphagia and frequent aspirations with ensuing bronchopulmonary infections are observed. Almost all patients with this type die within the first year of life.

In a few cases death already occurred shortly after birth, and only exceptional cases will die within the second year.

Radiological studies show the extreme biventricular enlargement of the heart (Fig. 4). The ECG typically shows a shortening of the PR interval, broadened QRS complexes, and signs of left- or biventricular hypertrophy. Changes of repolarisation can be found in about 10% of cases. Pre-excitation signs, like the Wolff-Parkinson-White syndrome, have repeatedly been reported [59, 61, 62, 63]. It has been suggested that the improvement of atrio-ventricular conduction is directly caused by glycogen-filled cells [64]. Unlike in constitutional activation of AMP kinase (see above) accessory atrioventricular pathways have never been described in patients with α-glucosidase deficiency.

Echocardiography again shows signs of left- or biventricular hypertrophy often associated with thickening of the intraventricular septum which may result in severe obstruction of the left-sided ventricular outflow tract. Only in rare cases is the right-sided outlow tract also affected [59, 61, 65]. The dynamics of the progressive disease process ultimately leading to severely impaired contractility have been described in detail [66, 67]. Only as a secondary phenomenon can dilatation and endocardial fibroelastosis with further restriction of contractility be

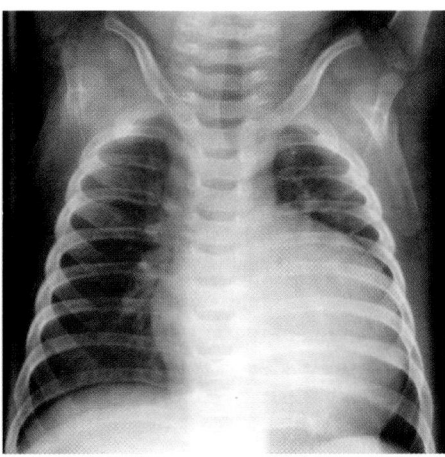

Fig. 4: Chest roentgenogram of the child from Figure 3 showing massive cardiomegaly with prominence the left ventricle.

observed [67, 68]; a pure dilated type of cardiomyopathy has not been described in α-glucosidase deficiency.

Pathoanatomically, the heart and skeletal muscle show marked vacuolation. This is the result of lysosomal glycogen storage which leads to the 'lace work appearance' of muscle tissue. To a much smaller extent, there may also be an increased cytosolic accumulation of glycogen [61]. No correlation could be found between the degree of glycogen storage and the residual glucosidase activity of the tissue. Autopsy findings have revealed that endocardial fibroelastosis, particularly of the left heart, occurs in about 20% of cases [61, 59, 69].

Fibroblast studies have suggested that α-glucosidase deficiency in lysosomes was more profound in early-onset than in late-onset cases [70]. The role of secondary factors cannot be excluded, however, because several adult patients were found with extremely low activity and almost no enzyme in the lysosomes [71]. Complementation studies yielded no sign of nonallelism of the several forms [70, 72].

Tab. 5: Clinicopathological features of GSD IIb (Danon's disease) *

	males	females
Cardiomyopathy	20 / 20	18 /18
Skeletal myopathy	2 / 20	6 /18
Mental retardation	13 / 16	1 /18
Clinical onset	< 20 years	adulthood

* from reference [98]

Currently, there is no curative treatment for this fatal disorder. However, several lines of research suggest the possibility of future treatment. Enzyme replacement strategies hold the greatest hope for patients currently affected by GSD II, but future strategies could include in vivo or ex vivo gene therapy approaches and/or mesenchymal stem cell or bone-marrow transplantation approaches. An interesting aspect is gene therapy in one tissue with transcellular effects for other tissues. It has recently been reported for a mouse model that muscle can act as a producer of α-glucosidase that is transferred to the heart where it was able to clear accumulated glycogen [73]. Thus, each of the approaches might eventually be combined to further improve the overall clinical efficacy of any one treatment regimen. Two preparations of recombinant human α-glucosidase are currently under investigation; one is an enzyme produced in Chinese hamster ovary cells while the other is produced in the milk of transgenic rabbits [74]. After successful experiments with human α-glucosidase in animal models of infantile α-glucosidase deficiency [75, 76], the first results of two studies using recombinant human α-glucosidase infused intravenously twice weekly in a small number of infants with infantile GSD IIa were reported in 2000 and 2001, respectively [77, 78]. In summary, these studies have demonstrated that a life-long intravenous enzyme replacement therapy might have the potential to minimize lysosomal glycogen accumulation. Both investigations showed that the enzyme preparation was generally well tolerated. Alpha-glucosidase activity normalized in tissue. Steady decreases in heart size and maintenance of normal cardiac function for more than 1 year were observed in these children. The infants lived well past the critical age of one year (16, 18, and 22 months at the time of one of the reports) and continued to have normal cardiac function. Improvements of skeletal muscle were also noted with marked improvement of muscle tone and strength. Furthermore, a normal neurological and developmental evaluation was reported [78]. On evaluation of morphological changes, not all infants responded equally well [79]. This was probably due to differences in the degree of glycogen storage and concomitant muscle pathology at the start of treatment. Animal studies further demonstrated that replacement of 20-30% of α-glucosidase is sufficient to clear glycogen in the heart of young mice while higher amounts are necessary for other tissues and for other age groups [80].

Lysosomal α-glucosidase deficiency, muscular type

Patients with α-glucosidase deficiency and an onset of symptoms not before toddler age, adolescence, or even adulthood have repeatedly been described [71, 72, 81, 82]. In these patients with later onset, symptoms are less progressive, and cardiomegaly, hepatomegaly, and macroglossia are not present. The clinical picture of this patient group is often characterized by a degenerative myopathy or by neurogenic muscular atrophy and myotonic manifestations, a clinical picture mimicking limb girdle dystrophy and bilateral paralysis of the diaphragm have been reported but also cases with aneurysms in the formation of cerebral arteries [83] have been published. Clinical signs usually start between 5 to 10 years of age, but the onset of symptoms, the clinical course, and the individual symptoms are extremely variable.

Therefore, diagnosis of this type of α-glucosidase deficiency is not easy. Plasma creatine kinase concentration which is generally elevated in the infantile form of GSD IIa [81, 84] has been reported to be normal in several patients with the late-onset form [59, 60]. Vacuolated lymphocytes can be demonstrated in the peripheral blood and this is independent of disease severity. Of lymphocytes 20 - 80 % may be vacuolated in lysosomal GSDs in contrast to <5 % in controls. The material within vacuoles is positive on PAS staining and is digested by diastase. Vacuolation is not increased in granulocytes [85]. Lysosomal storage can be visualized by electron microscopy of a skin biopsy. Deficiency of α-glucosidase can be performed in muscle tissue, fibroblasts, and lymphocytes. Because of the problems in differential diagnosis, most cases with the muscular form of GSD IIa will have a muscle biopsy where the storage phenomena in patients with the muscular type of GSD II are very differently pronounced in various muscle groups or even areas of one muscle group.

Although only very few cases of late-onset α-glucosidase deficiency involving the myocardium have been described [86], patients show an increased lysosomal glycogen storage in heart muscle. Why cardiomyopathy is rare in these cases has not yet been explained. Furthermore, myocardial involvement secondary to respiratory insufficiency with increased right ventricular mass has been observed. Like in the infantile cases, pre-excitation syndromes and atrioventricular heart blocks have been described for adult patients with the muscular type of α-glucosidase deficiency.

Treatment is mainly directed towards the neuromuscular symptoms. Since it is anticipated that α-glucosidase will be widely available for human use in the near future, also this patient group might benefit from life-long enzyme replacement therapy.

6.2.2.2 Lysosome associated membrane protein-2 deficiency, GSD IIb (Danon disease)

Repeatedly, cases with a vacuolar glycogen storage disease without any detectable defect of acid α-glucosidase have appeared in the literature [87,88,89,90]. Clinical features and histomorphological findings in such cases are very similar to GSD IIa with a variable age of presentation. Fatal neonatal cases have been reported [91,92] but also patients who reached adulthood. Death was reported generally to occur between ages 18 and 40 years. Also in these cases, muscle and heart muscle involvement are the most striking features. Many of these patients have severe mental retardation. Hepatomegaly was found in a relatively low percentage of cases (Table 5). Cases with glycogen storage limited to the heart have been published [93] and, like in GSD IIa, arrythmias and pre-excitation signs [94] as well as atrioventricular block [93] have been described.

Many pedigrees of affected families have suggested an X-chromosomal trait because both males and females are affected; however, females may have milder symptoms, their disease starts later and death usually is remarkably later than in males (Table 5) [95]. Furthermore, male-to-male transmission has never been reported in this condition. X-linked inheritance has finally been confirmed by the description of the basic defect. In 2000, it was reported that this condition is caused by mutations of a chromosomal gene coding for the lysosomal membrane-associated protein-2 (LAMP-2) [96]. Not in all cases of vacuolar glycogen storage without α-glucosidase deficiency, however, could LAMP-2 deficiency be detected and, therefore, it has been assumed that at least one other type of infantile autophagic vacuolar myopathy must exist [15, 97].

Apart from heart transplantation [95], no treatment has yet become available for patients with cardiomyopathy caused by LAMP-2 deficiency.

References

[1] Chen, Y. T.: Glycogen storage diseases. In: The metabolic and molecular bases of inherited disease. 8th ed., pp. 1521-1552. C.R. Scriver, A.L. Beaudet, W.S. Sly, D. Valle (eds.). McGraw-Hill Companies, New York, (2001).

[2] Shen, J. J., Y .T. Chen: Molecular characterization of glycogen storage disease type III. Curr. Mol. Med. **2**, 167-175 (2002).

[3] Shen, J., Y. Bao, H. M. Liu, P. Lee, J. V. Leonard, Y. T. Chen: Mutations in exon 3 of the glycogen debranching enzyme gene are associated with glycogen storage disease type III that is differentially expressed in liver and muscle. J. Clin. Invest. **98**, 352-357 (1996).

[4] Bao, Y, B. Z. Yang, T. L. Dawson Jr, Y. T. Chen: Isolation and nucleotide sequence of human liver glycogen debranching enzyme mRNA: identification of multiple tissue-specific isoforms. Gene **197**, 389-98 (1997).

[5] Gautron, S., D. Daegelen, F. Mennecier, D. Dubocq, A. Kahn, J.-C. Dreyfus: Molecular mechanisms of McArdle's disease (muscle glycogen phosphorylase deficiency). J. Clin. Invest. **79**, 275-281 (1987).

[6] Newgard, C. B., K. Nakano, P. K. Hwang, R. J. Fletterick: Sequence analysis of the cDNA encoding human liver glycogen phosphorylase reveals tissue-specific codon usage. Proc. Nat. Acad. Sci. **83**, 8132-8136 (1986).

[7] Mizuta, K., E. Hashimoto, A. Tsutou, Y. Eishi, T. Takemura, K. Narisawa, H. Yamamura: A new type of glycogen storage disease caused by deficiency of cardiac phosphorylase kinase. Biochem. Biophys. Res. Commun. **119**, 582-587 (1984).

[8] Eishi, Y., T. Takemura, R. Sone, H. Yamamura, K. Narisawa, R. Ichinohasama, M. Tanaka, S. Hatakeyama: Glycogen storage disease confined to the heart with deficient activity of cardiac phosphorylase kinase: a new type of glycogen storage disease. Hum. Pathol. **16**, 193-197 (1985).

[9] Servidei, S., L. A. Metlay, J. Chodosh, S. DiMauro: Fatal infantile cardiopathy caused by phosphorylase b kinase deficiency. J. Pediatr. **113**, 82-85 (1988).

[10] Elleder, M., Y. S. Shin, A. Zuntova, P. Vojtovic, V. Chalupecky: Fatal infantile hypertrophic cardiomyopathy secondary to deficiency of heart specific phosphorylase b kinase. Virchows Arch. A. Pathol. Anat. Histopathol. **423**, 303-307 (1993).

[11] Burchell, A., T. W. Cohen, P. Cohen: Distribution of isoenzymes of the glycogenolytic cascade in different types of muscle fibre. FEBS Lett. **67**, 17-22 (1976).

[12] Kilimann, M. W.: Molecular genetics of phosphorylase kinase: cDNA cloning, chromosomal mapping and isoform structure. J. Inherit. Metab. Dis. **13**, 435-441 (1990).

[13] Winchester, B. G.: Lysosomal membrane proteins. Eur. J. Paediatr. Neurol. **5** (suppl A), 11-19 (2001).

[14] Saftig, P., Y. Tanaka, R. Lüllmann-Rauch, K. von Figura: Disease model: LAMP-2 enlightens Danon disease. Trends Mol. Med. **7**, 37-39 (2001).

[15] Nishino, I.: Autophagic vacuolar myopathies. Curr. Neurol. Neurosci. Rep. **3**, 64-69 (2003).

[16] Steckel, F., V. Gieselmann, A. Waheed, A. Hasilik, K. von Figura, R. Oude Elferink, R. Kalsbeek, J. M. Tager: Biosynthesis of acid alpha-glucosidase in late-onset forms of glycogenosis type II (Pompe's disease). FEBS Lett. **150**, 69-76 (1982).

[17] Reuser, A. J., M. Kroos, R. P. Oude Elferink, J. M. Tager: Defects in synthesis, phosphorylation, and maturation of acid alpha-glucosidase in glycogenosis type II. J. Biol. Chem. **260**, 8336-8341 (1985).

[18] Reuser, A. J., M. Kroos, R. Willemsen, D. Swallow, J. M. Tager, H. Galjaard: Clinical diversity in glycogenosis type II. Biosynthesis and in situ localization of acid alpha-glucosidase in mutant fibroblasts. J. Clin. Invest. **79**, 1689-1699 (1987).

[19] Hermans, M. M., E. de Graaff, M. A. Kroos, H. A. Wisselaar, R. Willemsen, B. A. Oostra, A. J. Reuser: The conservative substitution Asp-645 → Glu in lysosomal alpha-glucosidase affects transport and phosphorylation of the enzyme in an adult patient with glycogen-storage disease type II. Biochem. J. **289**, 687-693 (1993).

[20] Hoefsloot, L. H., A. T. van der Ploeg, M. A. Kroos, M. Hoogeveen-Westerveld, B. A. Oostra, A. J. Reuser: Adult and infantile glycogenosis type II in one family, explained by allelic diversity. Am. J. Hum. Genet. **46**, 45-52 (1990).

[21] Brown, B.: Debranching and branching enzyme deficiencies. In: Myology, pp. 1585-1601. G. Engel, B. Q. Banker (eds.). McGraw-Hill Book Company, New York (1986).

[22] Coleman, R. A., H. S. Winter, B. Wolf, J. M. Gilchrist, Y. T. Chen: Glycogen storage disease type III (glycogen debranching enzyme deficiency): correlation of biochemical defects with myopathy and cardiomyopathy. Ann. Intern. Med. **116**, 896-900 (1992).

[23] Moses, S. W.: Muscle glycogenosis. J. Inherit. Metab. Dis. **13**, 452-465 (1990).

[24] Cornelio, F., N. Bresolin, P. A. Singer, S. DiMauro, L. P. Rowland: Clinical varieties of neuromuscular disease in debrancher deficiency. Arch. Neurol. **41**, 1027-1032 (1984).

[25] Labrune, P., P. Huguet, M. Odievre: Cardiomyopathy in glycogen-storage disease type III: clinical and echographic study of 18 patients. Pediatr. Cardiol. **12**, 161-163 (1991).
[26] Coleman, R. A., H. S. Winter, B. Wolf, Y. T. Chen: Glycogen debranching enzyme deficiency: long-term study of serum enzyme activities and clinical features. J. Inherit. Metab. Dis. **15**, 869-881 (1992).
[27] Moses, S. W., K. L. Wanderman, A. Myroz, M. Frydman: Cardiac involvement in glycogen storage disease type III. Eur. J. Pediatr. **148**, 764-766 (1989).
[28] Smit, G. P., J. Fernandes, J.V. Leonard, E.E. Matthews, S. W. Moses, M. Odievre, K. Ullrich: The long-term outcome of patients with glycogen storage diseases. J. Inherit. Metab. Dis. **13**, 411-418 (1990).
[29] Carvalho, J. S., E. E. Matthews, J. V. Leonard, J. Deanfield: Cardiomyopathy of glycogen storage disease type III. Heart Vessels **8**, 155-159 (1993).
[30] Talente, G. M., R. A. Coleman, C. Alter, L. Baker, B. I. Brown, R. A. Cannon, Y. T. Chen, J. F. Crigler Jr, P. Ferreira, J. C. Haworth, et al.: Glycogen storage disease in adults. Ann. Intern. Med. **120**, 218-226 (1994).
[31] Cuspidi, C., L. Sampieri, S. Pelizzoli, G. Pontiggia, A. Zanchetti, A. Nappo, V. Caputo, L. Matturri: Obstructive hypertrophic cardiomyopathy in type III glycogen-storage disease. Acta Cardiol. **52**, 117-123 (1997).
[32] Rossignol, A. M., M. Meyer, B. Rossignol, M. P. Palcoux, E. J. Raynaud, M. Bost: La myocardiopathie de la glycogenose type III. Arch. Fr. Pediatr **36**, 303-309 (1979).
[33] Lee, P. J., J. E. Deanfield, M. Burch, K. Baig, W. J. McKenna, J. V. Leonard: Comparison of the functional significance of left ventricular hypertrophy in hypertrophic cardiomyopathy and glycogenosis type III. Am. J. Cardiol. **79**, 834-838 (1997).
[34] Moon, J. C., H. R. Mundy, P. J. Lee, R. H. Mohiaddin, D. J. Pennell: Images in cardiovascular medicine. Myocardial fibrosis in glycogen storage disease type III. Circulation **107**, e47 (2003).
[35] Lee, P. J., C. Ferguson, F. W. Alexander: Symptomatic hyperinsulinism reversed by dietary manipulation in glycogenosis type III. J. Inherit. Metab. Dis. **20**, 612-613 (1997).
[36] Slonim, A. E., C. Weisberg, P. Benke, O. B. Evans, I. M. Burr: Reversal of debrancher deficiency myopathy by the use of high-protein nutrition. Ann. Neurol. **11**, 420-422 (1982).
[37] Schochet, S. S. jr., W. F. McCormick, H. Zellweger: Type IV glycogenosis (amylopectinosis). Light and electron microscopic observations. Arch. Pathol. **90**, 354-363 (1970).
[38] Ferguson, I. T., M. Mahon, W. J. Cumming: An adult case of Andersen's disease – Type IV glycogenosis. A clinical, histochemical, ultrastructural and biochemical study. J. Neurol. Sci. **60**, 337-351 (1983).
[39] Bruno, C., S. Servidei, S. Shanske, G. Karpati, S. Carpenter, D. McKee, R. J. Barohn, M. Hirano, Z. Rifai, S. DiMauro: Glycogen branching enzyme deficiency in adult polyglucosan body disease. Ann. Neurol. **33**, 88-93 (1993).
[40] Schröder, J. M., R. May, Y. S. Shin, M. Sigmund, S. Nase-Huppmeier: Juvenile hereditary polyglucosan body disease with complete branching enzyme deficiency (type IV glycogenosis). Acta Neuropathol. (Berl). **85**, 419-430 (1993).
[41] Servidei, S., R.E. Riepe, C. Langston, L.Y. Tani, J.T. Bricker, N. Crisp-Lindgren, H. Travers, D. Armstrong, S. DiMauro: Severe cardiopathy in branching enzyme deficiency. J. Pediatr. **111**, 51-66 (1987).
[42] Ferrans, V.J., R.G. Hibbs, J.J. Walsh, G.E. Burch: Cardiomyopathy, cirrhosis of the liver and deposits of a fibrillar polysaccharide. Report of a case with histochemical and electron microscopic studies. Am. J. Cardiol. **17**, 457-469 (1966).
[43] Craig, J. M., L. L. Uvman: A familial metabolic disorder with storage of an unusual polysaccharide complex. Pediatrics **22**, 20-32 (1958).
[44] Postler, E., E. Sindern, M. Vorgerd, I. Schmitz, J. P. Malin, K. M. Müller: Letale Kardiomyopathie bei adulter Polyglukosankörperkrankheit. Pathologe **23**, 229-234 (2002).
[45] Nase, S., K. P. Kunze, M. Sigmund, J. M. Schroeder, Y. Shin, P Hanrath: A new variant of type IV glycogenosis with primary cardiac manifestation and complete branching enzyme deficiency. In vivo detection by heart muscle biopsy. Eur. Heart J. **16**, 1698-1704 (1995).
[46] Tang, T. T., A. D. Segura, Y. T. Chen, L. M. Ricci, R. A. Franciosi, M. L. Splaingard, M. S. Lubinsky: Neonatal hypotonia and cardiomyopathy secondary to type IV glycogenosis. Acta Neuropathol. (Berl) **87**, 531-536 (1994).
[47] Starzl, T. E., A. J. Demetris, M. Trucco, C. Ricordi, S. Ildstad, P. I. Terasaki, N. Murase, R. S. Kendall, M. Kocova, W. A. Rudert, et al.: Chimerism after liver transplantation for type IV glycogen storage disease and type 1 Gaucher's disease. N. Engl. J. Med. **328**, 745-749 (1993).
[48] Selby, R., T. E. Starzl, E. Yunis, S. Todo, A. G. Tzakis, B. I. Brown, R. S. Kendall: Liver transplantation for type I and type IV glycogen storage disease. Eur. J. Pediatr. **152** (Suppl 1), S 71-76 (1993).
[49] Sokal, E. M., F. Van Hoof, D. Alberti, J. de Ville de Goyet, T. de Barsy, J. B. Otte: Progressive cardiac failure following orthotopic liver transplantation for type IV glycogenosis. Eur. J. Pediatr. **151**, 200-203 (1992).
[50] Ewert, R., A. Gulijew, R. Wensel, M. Dandel, M. Hummel, M. Vogel, R. Meyer, R. Hetzer: Die Glykogenose Typ IV als seltene Ursache einer Kardiomyopathie - Bericht einer erfolgreichen Herztransplantation. Z. Kardiol. **88**, 850-856 (1999).
[51] Amit, R., N. Bashan, J. M. Abarbanel, Y. Shapira, S. Sofer, S. Moses: Fatal familial infantile glycogen storage disease: multisystem phosphofructokinase deficiency. Muscle Nerve. **15**, 455-458 (1992).
[52] Burwinkel, B., S. Shiomi, A. Al Zaben, M. W. Kilimann: Liver glycogenosis due to phosphorylase kinase deficiency: PHKG2 gene structure and mutations associated with cirrhosis. Hum. Mol. Genet. **7**, 149-154 (1998).
[53] Burwinkel, B., A. J. Maichele, O. Aagenaes, H. D. Bakker, A. Lerner, Y. S. Shin, J.A. Strachan, M. W. Kilimann: Autosomal glycogenosis of liver and muscle due to phosphorylase kinase deficiency is caused by mutations in the phosphorylase kinase beta subunit (PHKB). Hum. Mol. Genet. **6**, 1109-1115 (1997).

[54] Arad, M., D. W. Benson, A. R. Perez-Atayde, W. J. McKenna, E. A. Sparks, R. J. Kanter, K. McGarry, J. G. Seidman, C. E. Seidman: Constitutively active AMP kinase mutations cause glycogen storage disease mimicking hypertrophic cardiomyopathy. J. Clin. Invest. **109**, 357-362 (2002).

[55] Gollob, M. H.: Glycogen storage disease as a unifying mechanism of disease in the PRKAG2 cardiac syndrome. Biochem. Soc. Trans. **31**, 228-231 (2003).

[56] Blair, E., C. Redwood, H. Ashrafian, M. Oliveira, J. Broxholme, B. Kerr, A. Salmon, I. Ostman-Smith, H. Watkins: Mutations in the gamma(2) subunit of AMP-activated protein kinase cause familial hypertrophic cardiomyopathy: evidence for the central role of energy compromise in disease pathogenesis. Hum. Mol. Genet. **10**, 1215-1220 (2001).

[57] Gollob, M. H., M. S. Green, A. S. Tang, T. Gollob, A. Karibe, A. S. Ali Hassan, F. Ahmad, R. Lozado, G. Shah, L. Fananapazir, L. L. Bachinski, R. Roberts, A. S. Hassan: Identification of a gene responsible for familial Wolff-Parkinson-White syndrome. N. Engl. J. Med. **344**, 1823-1831 (2001).

[58] Noori, S., R. Acherman, B. Siassi, C. Luna, M. Ebrahimi, Z. Pavlova, R. Ramanathan: A rare presentation of Pompe disease with massive hypertrophic cardiomyopathy at birth. J. Perinat. Med. **30**, 517-521 (2002).

[59] Ullrich, K.: Klinische und biochemische Variabilität der Glykogenose II (GSD II, M. Pompe). In: Stoffwechselerkrankungen im Kindesalter, pp. 148-161. K. Stehr, H.J. Böhles (eds.). Perimed Fachbuch-Verlagsgesellschaft, Erlangen (1987).

[60] Ullrich, K., H. Gröbe, R. Korinthenberg, D. B. von Bassewitz: Severe course of glycogen storage disease type II (Pompe's disease) without development of cardiomegalia. Pathol. Res. Pract., 627-632 (1986).

[61] Engel, A. G.: Acid maltase deficiency. In: Myology, pp. 1629-1651. A. G. Engel, B. Q. Banker (eds.). McGraw-Hill Book Company, New York (1986).

[62] Caddell, J. L., R. Whitemore: Observations on generalized glycogenosis with emphasis on electrocardiographic changes. Pediatrics **29**, 743-763 (1962).

[63] Bulkley, B.H., G.M. Hutchins: Pompe's disease presenting as hypertrophic myocardiopathy with Wolff-Parkinson-White syndrome. Am. Heart J. **96**, 246-252 (1978).

[64] Gillette, P. C., M. R. Nihill, D. B. Singer: Electrophysiological mechanism of the short PR interval in Pompe disease. Am. J. Dis. Child. **128**, 622-626 (1974).

[65] Rees, A., F. Elbl, K. Minhas, R. Solinger: Echocardiographic evidence of outflow tract obstruction in Pompe's disease (glycogen storage disease of the heart). Am. J. Cardiol. **37**, 1103-1106 (1976).

[66] Lam, J., L. J. Lubbers, M. S. J. Naeff, G. Losekoot: Glycogen storage disease type II (Pompe's disease): electrocardiographic and echocardiographic features. In: Pediatric Cardiology. Proceedings of the Second World Congress, pp. 1124-1125. E. G. Doyle, M. A. Engle, W. M. Gersooy, W. J. Rashkind, N. S. Talner (eds.). Springer Verlag, New York (1986).

[67] Seifert, B. L., M. S. Snyder, A. A. Klein, J. E. O'Loughlin, M. S. Magid, M. A. Engle: Development of obstruction to ventricular outflow and impairment of inflow in glycogen storage disease of the heart: serial echocardiographic studies from birth to death at 6 months. Am. Heart J. **123**, 239-242 (1992).

[68] Dickinson, D. F., W. T. Houlsby, J. L. Wilkinson: Unusual angiographic appearances of the left ventricle in 2 cases of Pompe's disease (glycogenosis type II). Br. Heart J. **41**, 238-240 (1979).

[69] Dincsoy, M. Y., H. P. Dincsoy, A. D. Kessler, M. A. Jackson, J. B. Sidbury jr.: Generalized glycogenosis and associated endocardial fibroelastosis. Report of 3 cases with biochemical studies. J. Pediatr. **67**, 728-740 (1965).

[70] Reuser, A. J., J. F. Koster, A. Hoogeveen, H. Galjaard: Biochemical, immunological, and cell genetic studies in glycogenosis type II. Am. J. Hum. Genet. **30**, 132-143 (1978).

[71] Reuser, A.J., M. Kroos, R. Willemsen, D. Swallow, J.M. Tager, H. Galjaard: Clinical diversity in glycogenosis type II. Biosynthesis and in situ localization of acid alpha-glucosidase in mutant fibroblasts. J. Clin. Invest. **79**, 1689-1699 (1987).

[72] Hoefsloot, L.H., A.T. van der Ploeg, M.A. Kroos, M. Hoogeveen-Westerveld, B.A. Oostra, A.J. Reuser: Adult and infantile glycogenosis type II in one family, explained by allelic diversity. Am. J. Hum. Genet. **46**, 45-52 (1990).

[73] Martin-Touaux, E., J. P. Puech, D. Chateau, C. Emiliani, E. J. Kremer, N. Raben, B. Tancini, A. Orlacchio, A. Kahn, L. Poenaru: Muscle as a putative producer of acid alpha-glucosidase for glycogenosis type II gene therapy. Hum. Mol. Genet. **11**, 1637-1645 (2002).

[74] Reuser, A. J., H. Van Den Hout, A. G. Bijvoet, M. A. Kroos, M. P. Verbeet, A. T. Van Der Ploeg: Enzyme therapy for Pompe disease: from science to industrial enterprise. Eur. J. Pediatr. **161** (Suppl 1), 106-111 (2002).

[75] Bijvoet, A. G., H. Van Hirtum, M. A. Kroos, E. H. Van de Kamp, O. Schoneveld, P. Visser, J. P. Brakenhoff, M. Weggeman, E. J. van Corven, A.T. Van der Ploeg, A. J. Reuser: Human acid alpha-glucosidase from rabbit milk has therapeutic effect in mice with glycogen storage disease type II. Hum. Mol. Genet. **8**, 2145-2153 (1999).

[76] Kikuchi, T., H. W. Yang, M. Pennybacker, N. Ichihara, M. Mizutani, J. L. Van Hove, Y. T. Chen: Clinical and metabolic correction of Pompe disease by enzyme therapy in acid maltase-deficient quail. J. Clin. Invest **101**, 827-833 (1998).

[77] Van den Hout, H., A. J. Reuser, A. G. Vulto, M. C. Loonen, A. Cromme-Dijkhuis, A. T. Van der Ploeg: Recombinant human alpha-glucosidase from rabbit milk in Pompe patients. Lancet **356**, 397-398 (2000).

[78] Amalfitano, A., A. R. Bengur, R. P. Morse, J. M. Majure, L. E. Case, D. L. Veerling, J. Mackey, P. Kishnani, W. Smith, A. McVie-Wylie, J. A. Sullivan, G. E. Hoganson, J.A. Phillips 3rd, G. B. Schaefer, J. Charrow, R. E. Ware, E. H. Bossen, Y. T. Chen: Recombinant human acid alpha-glucosidase enzyme therapy for infantile glycogen storage disease type II: results of a phase I/II clinical trial. Genet. Med. **3**, 132-138 (2001).

[79] Winkel, L. P., J. H. Kamphoven, H. J. van den Hout, L. A. Severijnen, P. A. van Doorn, A. J. Reuser, A. T. van der Ploeg: Morphological changes in muscle tissue of

patients with infantile Pompe's disease receiving enzyme replacement therapy. Muscle Nerve **27**, 743-751 (2003).

[80] Raben, N., T. Jatkar, A. Lee, N. Lu, S. Dwivedi, K. Nagaraju, P. H. Plotz: Glycogen stored in skeletal but not in cardiac muscle in acid alpha-glucosidase mutant (Pompe) mice is highly resistant to transgene-encoded human enzyme. Mol. Ther. **6**, 601-608 (2002).

[81] Ausems, M. G., P. Lochman, O. P. van Diggelen, H. K. Ploos van Amstel, A. J. Reuser, J. H. Wokke: A diagnostic protocol for adult-onset glycogen storage disease type II. Neurology **52**, 851-853 (1999).

[82] Matsuishi, T., M. Yoshino, K. Terasawa, I. Nonaka: Childhood acid maltase deficiency. A clinical, biochemical, and morphologic study of three patients. Arch. Neurol. **41**, 47-52 (1984).

[83] Makos, M. M., R. D. McComb, M. N. Hart, D. R. Bennett: Alpha-glucosidase deficiency and basilar artery aneurysm: report of a sibship. Ann. Neurol. **22**, 629-633 (1987).

[84] van den Hout, H. M., W. Hop, O. P. van Diggelen, J. A. Smeitink, G. P. Smit, B. T. Poll-The, H. D. Bakker, M. C. Loonen, J. B. de Klerk, A. J. Reuser, A. T. van der Ploeg: The natural course of infantile Pompe's disease: 20 original cases compared with 133 cases from the literature. Pediatrics **112**, 332-340 (2003).

[85] von Bassewitz, D. B., H. Gröbe, K. Ullrich: Atypische klinische Formen der Typ II Glycogenose (Pompe). Verh. Dtsch. Ges. Pathol. **66**, 319-323 (1982).

[86] Suzuki, Y., A. Tsuji, K. Omura, G. Nakamura, S. Awa, M. Kroos, A. J. Reuser: Km mutant of acid alpha-glucosidase in a case of cardiomyopathy without signs of skeletal muscle involvement. Clin. Genet. **33**, 376-385 (1988).

[87] Danon, M. J., S. J. Oh, S. DiMauro, J. R. Manaligod, A. Eastwood, S. Naidu, L. H. Schliselfeld: Lysosomal glycogen storage disease with normal acid maltase. Neurology **31**, 51-57 (1981).

[88] Antopol, W., E. P. Boas, W. Levison, L. R. Tuchman: Cardiac hypertrophy caused by glycogen storage disease in a 15-year-old boy. Am. Heart J. **20**, 546-556 (1940).

[89] Byrne, E., X. Dennett, B. Crotty, I. Trounce, J. M. Sands, R. Hawkins, J. Hammond, S. Anderson, E. A. Haan, A. Pollard: Dominantly inherited cardioskeletal myopathy with lysosomal glycogen storage and normal acid maltase levels. Brain **109**, 523-536 (1986).

[90] Tachi, N., M. Tachi, K. Sasaki, H. Tomita, S. Wakai, S. Annaka, R. Minami, S. Tsurui, H. Sugie: Glycogen storage disease with normal acid maltase: skeletal and cardiac muscles. Pediatr. Neurol. **5**, 60-63 (1989).

[91] Atkin, J., J. W. Snow jr., H. Zellweger, W. J. Rhead: Fatal infantile cardiac glycogenosis without acid maltase deficiency presenting as congenital hydrops. Eur. J. Pediatr. **142**, 150 (1984).

[92] Verloes, A., J. M. Massin, J. Lombet, B. Grattagliano, D. Soyeur, J. Rigo, L. Koulischer, F. Van Hoof: Nosology of lysosomal glycogen storage diseases without in vitro acid maltase deficiency. Delineation of a neonatal form. Am. J. Med. Genet. **72**, 135-142 (1997).

[93] Tripathy, D., R. A. Coleman, H. J. Vidaillet jr., C. Steenbergen, K. Hirschhorn, D. L. Packer: Complete heart block with myocardial membrane-bound glycogen and normal peripheral alpha-glucosidase activity. Ann. Intern. Med. **109**, 985-987 (1988).

[94] Riggs, J. E., S. S. Schochet jr., L. Gutmann, S Shanske, W. A. Neal, S. DiMauro: Lysosomal glycogen storage disease without acid maltase deficiency. Neurology **33**, 873-877 (1983).

[95] Dworzak, F., F. Casazza, M. Mora, R. De Maria, E. Gronda, G. Baroldi, M. Rimoldi, L. Morandi, F. Cornelio: Lysosomal glycogen storage with normal acid maltase: a familial study with successful heart transplant. Neuromuscul. Disord. **4**, 243-247 (1994).

[96] Nishino, I., J. Fu, K. Tanji, T. Yamada, S. Shimojo, T. Koori, M. Mora, J. E. Riggs, S. J. Oh, Y. Koga, C. M. Sue, A. Yamamoto, N. Murakami, S. Shanske, E. Byrne, E. Bonilla, I. Nonaka, S. DiMauro, M. Hirano: Primary LAMP-2 deficiency causes X-linked vacuolar cardiomyopathy and myopathy (Danon disease). Nature **406**, 906-910 (2000).

[97] Yamamoto, A., Y. Morisawa, A. Verloes, N. Murakami, M. Hirano, I. Nonaka, I. Nishino: Infantile autophagic vacuolar myopathy is distinct from Danon disease. Neurology **57**, 903-905 (2001).

[98] Sugie, K., A. Yamamoto, K. Murayama, S. J. Oh, M. Takahashi, M. Mora, J. E. Riggs, J. Colomer, C. Iturriaga, A. Meloni, C. Lamperti, S. Saitoh, E. Byrne, S. DiMauro, I. Nonaka, M. Hirano, I. Nishino: Clinicopathological features of genetically confirmed Danon disease. Neurology **58**, 1773-1778 (2002).

7 Cardiomyopathies and mitochondrial defects of oxidative energy metabolism

W. Sperl

During the last 20 years a variety of very heterogeneous defects of mitochondrial energy production have been characterized; the clinical presentation being dominated by neuromuscular disturbances with frequent systemic involvement of the entire organism. Since organs with a high oxygen and energy requirement are mainly affected, it is not astonishing that cardiac involvement has frequently been described; however, characterization of the cardiac defects has lagged considerably behind the neuromuscular mitochondrial disease entities. There are some reasons for the late investigation of the cardiac mitochondrial energy metabolism in cardiomyopathies. In many mitochondrial disorders an exclusive cardiac involvement is rare and often the neuromuscular symptoms dominate. In addition, when compared to skeletal muscle, a sufficient amount of heart muscle tissue is much more difficult to obtain for biochemical investigations.

After Wallace et al. [122] and Holt et al. [43] in 1988 were the first to describe independently mtDNA mutations as the molecular basis of disease, a large number of descriptions of mitochondrial mtDNA abnormalities followed in cases of cardiomyopathy. The PCR technique made the analysis of extremely minute tissue samples possible. In addition, biochemical [91] and immunohistochemical [70, 71] methods of analysis and respirometric determinations in permeabilized muscle fibres for characterization of mitochondrial energy metabolism in extremely small tissue samples have been developed [56, 104, 120].

The discovery of disturbances of the oxidative phosphorylating (OXPHOS) system offered important new aspects for the understanding of cardiomyopathies [53, 55, 84]. Besides viral infection, alcohol, pregnancy, microvascular hyperreactivity, autoimmunity, free radical action, nuclear DNA mutations [49], defects of the mitochondrial energy generating system have been added to the list of possible causes of cardiomyopathies.

The present knowledge of OXPHOS system changes in different cardiomyopathies is confounding and a systematic classification is hardly possible. It is the objective of this article to review the cardiac changes in defects of OXPHOS metabolism and to show their enormous clinical heterogeneity on the basis of the available literature.

7.1 Characteristics of the oxidative phosphorylating system

The respiratory chain (Fig. 1) is the ultimate common pathway of substrate oxidation. The respiratory chain proteins are coded by the cellular nucleus as well as by mitochondrial DNA. Thus a complicated interaction exists between nuclear and mitochondrial DNA. Mitochondrial DNA contains 16 569 base pairs, has a circular structure and encodes 13 hydrophobic proteins of the mitochondrial electron transport system [7 subunits of complex I, apocytochrome b of complex III, 3 subunits of complex IV) including ATPase [2 subunits) (Fig. 2). As mitochondria are exclusively donated from the ovary during zygote formation, mtDNA mutations present a maternal pattern of inheritance. During this event a random distribution of mutated and wild-type DNA material (heteroplasmy) appears in the respective daughter cells with the result that at the end of the embryonic differentiation, defective DNA may be unevenly distributed between different tissues (tissue specificity of mitochondrial defects [45, 71, 103]). The phenotype of the individual OXPHOS defect is dependent on various factors: biochemical type of defect and tissue distribution, the individual oxygen requirement ("threshold effect") of the affected tissue [97, 123] and finally patient age and gender. On the grounds of these characteristics, the enormous clinical heterogeneity in mitochondrial defects may be better understood [27,

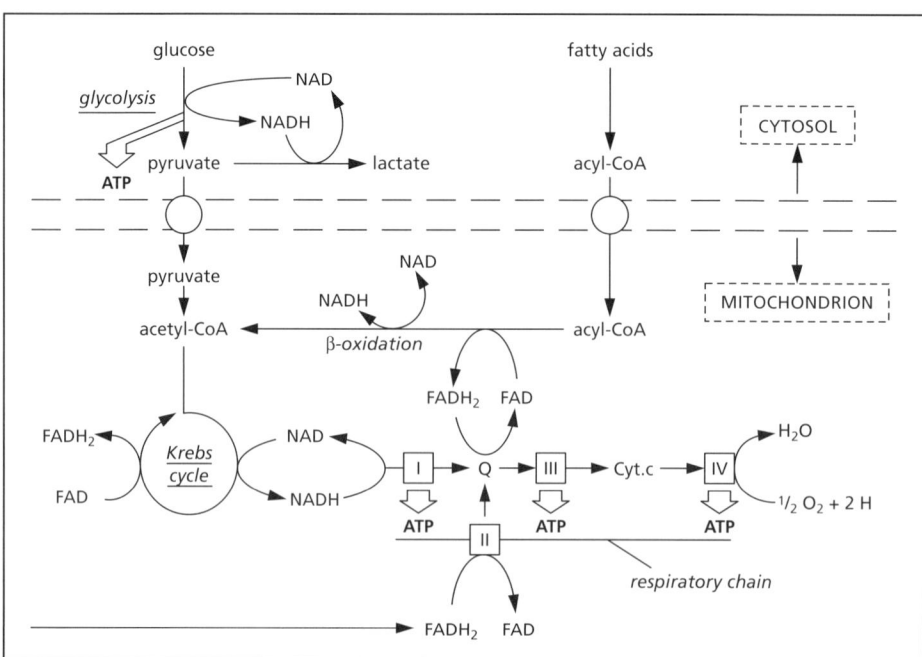

Fig. 1: Schematic representation of the mitochondrial respiratory chain in relation to part of the intermediary metabolism (from [124] with permission).

29, 74]. As a whole, different forms of OXPHOS system disturbances can be distinguished:
– A. Aging phenomena
– B. Secondary disturbances
– C. Primary deficiencies

7.2 Aging presbycardia

Accumulation of mitochondrial genome mutations during life is an important contributor to aging and degenerative diseases [61], especially in neuromuscular disorders and cardiac disease [39, 40, 70]. After the 40th year of life, mtDNA deletions have been regularly described [17] in heart muscle, which accumulate exponentially with increasing age (Fig. 3) [22, 23, 39, 40, 84]. This accumulation leads finally to non atherosclerotic heart dysfunction (= presbycardia). A frequent 4977 base pair deletion (mtDNA^{4977}) has been described [22, 39]; however, also mtDNA^{7436} and mtDNA^{10422} deletions were observed [23]. The 7.4 kbp deletion occurs commonly among mitochondrial cardiomyopathy patients [83] and also in normal elderly subjects [38]. In coronary sclerosis, the same deletions are significantly more frequent than in age-matched controls [22, 39]. Also histochemical studies have revealed aging in human heart muscle. Müller-Höcker

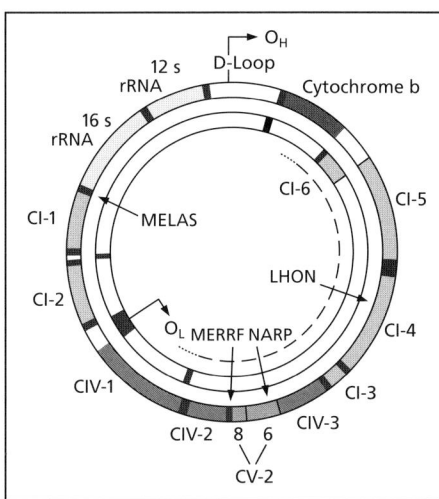

Fig. 2: Map of human mtDNA. CI-1 to CI-6 are genes for subunits 1 to 6 of complex I. Cytochrome b is the gene for the cytochrome b subunit of complex III. CIV-1 to CIV-3 are the genes for subunits 3 to 6 of complex IV. CV-6 and CV-8 are genes for subunits 6 and 8 of complex V(=ATPase). 16s RNA and 12s RNA are genes for ribosomal RNA. The D-loop is the control site for both transcription and replication. OH and OL are starting points of heavy and light strand replicationj. The small shaded areas dispersed through the mitochondrial genome are the 22 tRNA genes. The dashed circle-part on the inside of the mtDNA represents the area of the most common mtDNA deletions. MERRF, LHON, NARP and MELAS indicate the (most common) point mutations seen in these syndromes. From [124] with permission.

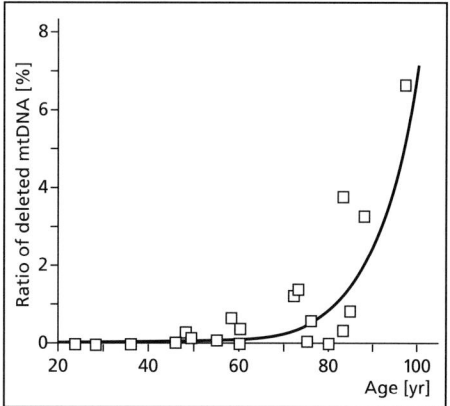

Fig. 3: The ratio of deleted mt DNA (7.4 kb-deletion) to the total mt DNA increases with age of the subjects (from 40).

demonstrated an increase in focal deficit of cytochrome oxidase activity [70]. In human mt DNA, 8-OH-deoxyguanosine formation from deoxyguanosine by active oxygen can be seen as a marker of oxidative damage [40]. Hayakawa [40] could show that an increase in mt DNA 8-OH-deoxyguanosine content is closely correlated with the age dependent increase in deletions. Also in animal studies, the involvement of the OXPHOS system in aging has been proven. In the rat heart, decreased complex I and IV activities were demonstrated [40], F_0F_1ATP-synthase activity [36] and the mtDNA content [5] were decreased with increasing age.

7.3 Cardiomyopathies as a consequence of a secondary damage of the OXPHOS system (secondary mitochondrial cardiomyopathy)

As a consequence of its close localisation near to the respiratory chain within mitochondria, mtDNA is particularly vulnerable. Electron flow causes the generation of oxygen radicals that damage the mtDNA and may lead to mutations, because protective histones as well as a corresponding repair system are lacking. In addition, the high replication frequency of mt DNA causes a 5 – 10 times increased rate of spontaneous mutations. Secondary disturbances of the OXPHOS system caused by radical formation may occur in any case of ischaemia and hypoxia and can also affect cardiac muscle [22, 23, 58, 67]. A disturbed energy metabolism in heart muscle can also be caused by medications like zidovudine [57, 60] and doxorubicine [1, 19] as well as by alcohol [26] and during allograft rejection [30]. Keshan disease is an exemplary secondary mitochondrial cardiomyopathy on the basis of an impaired mitochondrial anti-oxidative capacity caused by a decreased selenium content in combination with decreased respiratory chain enzyme activities [126]. As an adverse effect of zidovudine therapy, cardiomyopathy may also be observed in addition to the known myopathy. In these cases a depletion of mtDNA has also been observed [57]. The cardiac toxicity of doxorubicine as well as of other anthracyclins, when the cumulative dosage exceeds 450 mg/m^2 body surface area, is also considered to be based on radical damage to the OXPHOS system [101]. Radical formation leads to lipid peroxidation and secondary mtDNA damage [1].

It is important to note that secondary defects and physiological mtDNA mutations, which occur during aging, can add up in the respective tissues [108]. If finally a certain threshold of accumulation of defective DNA is passed, OXPHOS capacity decreases in the affected tissue. The consequences are functional disturbances of cellular functions, organ dysfunction and in patients a variety of clinical symptoms including death [123].

Contradictory opinions exist with regard to the role of OXPHOS in inborn heart malformations and coronary sclerosis compared with primary cardiomyopathies. In these cases different results have been obtained with different methods [16, 66]. Maurer and Zierz found low activities of heart muscle respiratory chain enzymes in patients with coronary sclerosis and heart insufficiency in contrast to patients with idiopathic dilative cardiomyopathy and aortic stenosis [66].

In the latter cases even a compensatory stimulation of the respiratory chain could be present. Buchwald et al. in contrast demonstrated a decrease of cytochromes and respiratory chain enzymes in dilative cardiomyopathies [16]. In both investigations, however, no functional respirometric determinations in intact mitochondria were performed.

7.4 Primary mitochondrial cardiomyopathies

A mitochondrial cardiomyopathy can be defined as a disease characterized by structurally, numerically, or functionally abnormal mitochondria, or a combination of these [63]. The definition of primary mitochondrial cardiomyopathy is difficult. It can be done on the basis of mt DNA mutations alone [84]. Nevertheless there are a number of reports on various cardiomyopathies with disturbances of the OXPHOS system characterized only at the enzymatic or immunohistochemical level where the defect is causally related to the disease.

For example, histiocytoid cardiomyopathy, a paediatric speciality (also termed arachnocytosis of the heart muscle, isolated lipoidosis of the heart, focal lipid cardiomyopathy, infantile cardiomyopathy with histiocytoid changes or oncocytary cardiomyopathy) [12, 31, 85, 98, 121] was the first cardiomyopathy in which an enzyme defect of the OXPHOS system was found (Figs. 4-6). Histology alone, showing abnormal mitochondria as well as lipid and glycogen storage phenomena, was suspicious for a disturbance of the OXPHOS system (Figs. 5,6). Finally a defect of com-

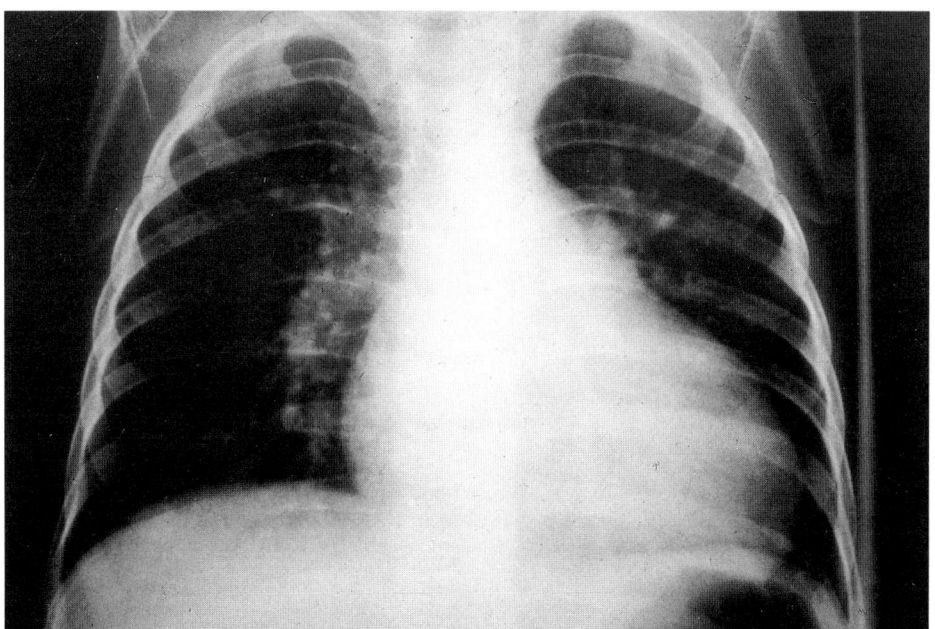

Fig. 4: Marked ventricular enlargement of a one year old patient with histiocytoid cardiomyopathy (12) and a decreased cytochrome aa3 content in heart and skeletal muscle.

plex III [85] or of cytochrome C oxidase (complex IV) [12] was demonstrated in isolated heart muscle mitochondria. These biochemical findings can sufficiently explain the morphological abnormalities justifying the term of primary mitochondrial cardiomyopathy.

Furthermore the classification of primary cardiomyopathy can be extremely difficult, because aging phenomena as well as secondary disturbances are not easy to separate from primary genetic defects.

A cardiac affection in mitochondriopathies is frequently described, however rarely dominates the clinical picture [99]; hence the disease can present as a fatal infantile cardiomyopathy [6, 49, 76]. In some infants with this fatal and early cardiomyopathy enzyme defects could be demonstrated (Table 1), in others not, despite the suggestive presentation as mitochondrial dysfunction [102]. In total, only in a minority of cases has heart muscle tissue been enzymatically studied when mitochondrial cardiomyopathies were described [12, 32, 111, 129, 131]. In many cases the defect was demonstrated in another tissue, frequently in skeletal muscle [6, 12, 29, 32, 71, 73, 85, 97, 114, 127]. The distribution frequency between affected heart and skeletal muscle is hitherto totally unknown.

It has to be emphasized that in all classical forms of cardiomyopathies [19], such as histiocytoid cardiomyopathy [12, 31, 85], dilative cardiomyopathy [6, 14, 20, 49, 54, 80, 106], hypertrophic cardiomyopathy [32, 44, 49, 83, 86, 92, 93, 95, 118] and even in cardiomyopathy with endocardial fibroelastosis [6, 41, 47, 82] OXPHOS alterations could be demonstrated either histologically or

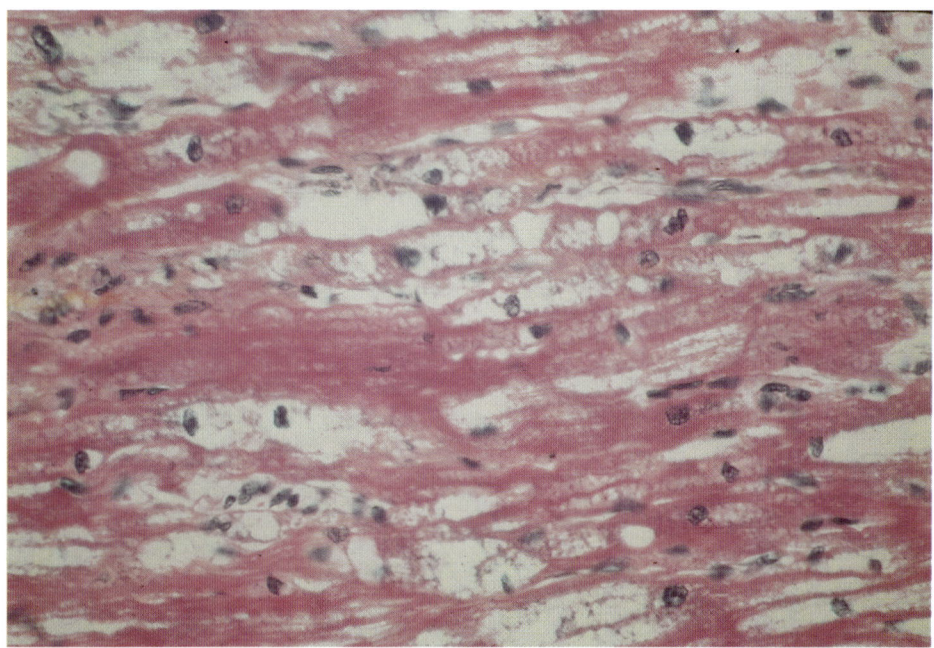

Fig. 5: Histology of heart muscle reveals apparently empty myocardial cells with foamy appearance.

enzymatically or, as in most cases, at the DNA level. At the present level of knowledge a suggested classification of mitochondrial cardiomyopathies can only be descriptive. In presenting different suggestions, an attempt is now made to examine them from different view points.

7.5 Classification of mitochondrial cardiomyopathy

7.5.1 Inheritance

Similar to the biochemical and clinical heterogeneity, a variable pattern of inheritance is equally found in cases of mitochondrial cardiomyopathies.

7.5.1.1 Maternally inherited cardiomyopathy

Maternal inheritance is a hallmark of mt DNA mutation and has been described in cases with infantile fatal cardiomyopathy due to point mutations of mt DNA [99, 130].

7.5.1.2 X-linked, recessive dilative cardiomyopathy

Of all dilated cardiomyopathies, some of them follow an X-linked inheritance. The genesis of X-linked dilated cardiomyopathy is still unknown. Some at least show abnormal mitochondria [82] and enzymatic defects of the respiratory chain [6].

Fatal with systemic disease: Abnormal mitochondria were found in heart, skeletal muscle, liver and kidney [76].

Fig. 6: Electron microscopy of the heart. There is a loss and disarray of myofibrils and an increased amount of glycogen. Empty appearance on light microscopy (Fig. 4) was due to accumulation of mitochondria.

With affected bone marrow (=Barth syndrome – BTHS, 6): Structural mitochondrial abnormalities were found in cardiac muscle, neutrophil bone marrow cells and occasionally in skeletal muscle [6]. Ino et al. described 2 boys with cardiomyopathy, neutropenia and short stature [47]. In addition 3-methylglutaconic aciduria can be found (= methylglutaconic aciduria type II) [107]. There is an association with isolated left ventricular noncompaction (spongy myocardium, persistent myocardial sinus, arrest in endomyocardial morphogenesis, excessive prominent trabeculations) [11]. The gene responsible for Barth syndrome has been mapped to Xq28 [2, 14], the gene is the G4.5 gene encoding for taffazines [10, 25]. Only one report on therapy with pantothenic acid exists [81]. It is important to note that neutropenia is variable and 3-methylglutaconic acid levels can fluctuate.

With 3-methylglutaconic acid excretion (= methylglutaconic aciduria type IV): This is a heterogeneous unclassified group of disorders [107]. Cardiomyopathy occurs together with muscular hypotonia, retardation, seizures, dysmorphic features, in some patients with lactic acidosis. A benign clinical presentation is possible [52]. 3-methylglutaconic acid excretion has also been found by other authors describing cardiomyopathy with OXPHOS defects [32, 46].

With endocardial fibroelastosis [41, 82].

7.5.1.3 Autosomal recessive hypertrophic cardiomyopathy with cataract (Sengers syndrome)

An autosomal recessively inherited syndrome characterized by mitochondrial myopathy of both heart and skeletal muscle, congenital cataract and lactic acidosis was first described by Sengers et al. in 1975 [93, 95]. The metabolic cause of the disease is so far not known. Biochemical investigation of mitochondrial function showed no abnormalities [93, 102].

An association with adenine nucleotide translocator defect has to be elucidated [51].

Mitochondrial hypertrophic cardiomyopathy should not be confused with the autosomal dominant hypertrophic cardiomyopathy [118], also described as an autosomal recessive trait (chromosome 14 q1) [50]. Based on data of a greater number of patients with Sengers syndrome, two different forms can be distinguished [119]:

Fatal neonatal form: Bilateral cataracts, lactacidaemia and mitochondrial myopathy are present from birth [118]. Hypertrophic obstructive cardiomyopathy is recognised within days after birth and causes death within weeks due to early left ventricular outflow obstruction.

Benign form: Hypertrophic cardiomyopathy and outflow obstruction develops later. Cataract may be the only symptom during early childhood [119].

7.5.1.4 Non-autosomal dominant hypertrophic cardiomyopathy and mitochondrial DNA mutation

Multiple point mutations have been described in association with non-autosomal dominant hypertrophic cardiomyopathy [79].

7.5.1.5 Inherited idiopathic dilative cardiomyopathy

Dilative cardiomyopathy is the most frequent severe cardiomyopathy in juveniles and young adults (3.65/10000]. Of these cases, 20% are familiar. Some show an X-chromosomal inheritance [7], others an autosomal recessive or dominant inheritance.

In inherited idiopathic dilatative cardiomyopathy, multiple mt DNA deletions [83, 106] have been described. It has been discussed that the deletions were not inherited as such but rather that an unknown nuclear gene defect leads to multiple mtDNA deletions [106].

7.5.2 Enzyme defects

Defects of the respiratory chain:
- complex I [13, 27, 69, 109]
- complex II [4]
- complex III [85]
- complex IV [32, 72, 80, 89, 94, 129]
- complex V (ATP-synthase) associated with methylglutaconic aciduria [42]
- combined: complex I and IV [71, 75, 90, 109, 111, 114, 131]
- multiple respiratory chain defects with 3-methylglutaconic acid excretion [32, 46]

Defects of the respiratory chain are associated in most cases with a hypertrophic cardiomyopathy. However, histiocytoid [85] and dilative cardiomyopathic forms have also been described [6, 16, 112]. It is surprising that cardiomyopathy has not yet been described in defects of the pyruvate dehydrogenase complex. Only in animal studies in the cardiomyopathic Syrian hamster has pyruvate dehydrogenase been found to be present in a less active state in the glucose perfused heart [28].

7.5.3 Characterization of the gene defect

The respiratory chain is regulated at the nuclear as well as at the mt DNA level. In cardiomyopathies, defects in both genomes are possible [84].

7.5.3.1 Nuclear DNA defects

A large number of nuclear OXPHOS genes exists. Nuclear mutations in mitochondrial proteins as well as mutations in nuclear genes involved in the synthesis, regulation, processing, delivery or translocation of nuclear gene products encoding mitochondrial components can lead to a disturbance in the OXPHOS system. Nuclear OXPHOS mutations are mainly recognized through indirect criteria [97]. A patient with lethal infantile mitochondrial disease (LIMD) and a possible nuclear DNA mutation was described by Zheng et al. [131]. Lethal infantile cardiomyopathy (LIC) [6, 49, 76], benign infantile myopathy and cardiomyopathy BIMC [13], X-linked hypertrophic cardiomyopathy [82] as well as some patients with pedigrees of autosomal dominant transmission with Kearns-Sayre/chronic external ophthalmoplegia plus syndromes (KS/CPEO) are candidates for nuclear DNA mutations [97].

7.5.3.2 mt DNA defects

Point mutations have been described in pure myopathy/cardiomyopathy [49, 79, 99, 100, 130] as well as in cardiomyopathies in association with mitochondrial multisystemic disease [112] or syndromes such as MELAS [49, 84, 111], MERRF [96] and Leigh syndrome [86]. Whereas in MELAS (A-G transition at nucleotide 3243) or MERRF syndrome

(A-G at 8344) characteristic single point mutations exist, multiple point mutations are described in most mitochondrial cardiomyopathies [84].

Deletions: a 7.4 kbp deletion is found in patients with mitochondrial cardiomyopathy [83] as well as in presbycardia [39]. Multiple deletions have been described in inherited idiopathic dilated cardiomyopathy [106] as well as in dilative [38, 49] and hypertrophic cardiomyopathies [49, 84]. Deletions have been also demonstrated in cardiomypopathies associated with Kearns-Sayre syndrome [3, 34, 72, 88]. mt DNA depletion has also been described with hypertrophic cardiomyopathy [64].

7.5.4 Mitochondrial syndromes with heart involvement

Several diseases with mitochondrial defects have been classified as syndromes according to a typical constellation of symptoms [29]. Some of them show cardiac involvement:

Cardiomyopathies have been described in Sengers syndrome [93, 119], Barth syndrome [6], Leigh syndrome [24, 86, 92, 97], chronic progressive external ophthalmoplegia (CPEO) [45] and Kearns-Sayre syndrome [3, 18, 34, 54, 72, 88, 128], MELAS (Mitochondrial encephalo-pathy, lactic acidosis, "stroke-like episodes") [48, 73, 111] and MERRF syndrome [96] and lethal infantile mitochondrial disease (LIMD) [75]. Together with Friedreich ataxia [105] some of these disorders can be classified as a group of "cardioneurological syndromes" [92].

Disturbances of cardiac rhythm are associated with CPEO [45], Kearns-Sayre syndrome [3, 34, 54, 72, 88], Leber`s hereditary neuropathy of the optic nerve (LHON) [77, 78] and MERRF syndrome (myoclonus epilepsy, "ragged red fibres") [128].

7.6 Disturbances of cardiac rhythm in defects of the OXPHOS system

In mitochondrial defects, the cardiac conduction system can also be involved. It seems to be that cells of the cardiac conduction system are highly vulnerable to a diminished energy supply.

Supraventricular conduction defects: Preexcitation syndromes such as LGL- and WPW-syndrome have been described especially in LHON [77, 78, 97]. They can be found also in mitochondrial myopathy with complex I defect [9], in oncocytic cardiomyopathy [98] and maternally inherited myopathy and cardiomyopathy [130]. Supraventricular tachycardia can be associated with CPEO [45].

Heart block: In Kearns-Sayre syndrome, cardiac anomalies are characteristically associated with the conduction system. The most common conduction abnormality is left anterior hemiblock, alone or in combination with a right bundle branch block. Others include Mobitz type II second degree atrioventicular and complete heart block [88]. Advanced atrioventricular block was also described in a 37-year-old woman with scapulohumeral muscular atrophy, rigid spine and cardiopathy and mitochondrial abnormalities [110].

Ventricular arrhythmias: In a 69-year-old patient ventricular arrythmias led to the discovery of mitochondrial myopathy [20]. Remarkable is the late onset of the disorder and the fact that clinical signs were exclusively cardiac.

7.7 Frequency of cardiac involvement in mitochondrial (encephalo-) myopathies
(see also [68])

Tulinius [115, 116] recently described 25 children with a well characterized OXPHOS defect and mitochondrial encephalomyopathy, of whom 12 presented cardiac involvement: 4 with hypertrophic cardiomyopathy, 3 with conduction defects, 4 with preexcitatory disturbances and one with disturbances of the sinus rhythm.

Böhles [12] described one female patient with histiocytoid cardiomyopathy and cytochrome C oxidase (COX) deficiency and reviewed among 16 patients with COX defects, 3 with suspected cardiomyopathy.

22 patients with mitochondrial encephalomyopathy have been diagnosed during the last 3 years at the Children`s Hospital, University of Innsbruck. Two of them presented hypertrophic cardiomyopathy in addition to methylglutaconic acid excretion. One child had WPW syndrome and three a partial right branch bundle block with a disturbance of repolarisation.

In a review of the literature for mitochondrial encephalomyopathies and their clinical manifestation (unpublished data), 143 reports of mitochondrial encephalomyopathies between 1987 and 1991 were investigated. In 12 cases a cardiomyopathy was reported: 9 hypertrophic cardiomyopathies, 4 arrhythmias including one WPW syndrome. It has to be emphasized that the cardiac involvement was exclusively detected during a special cardiological examination. Not in all cases could histology be obtained.

In conclusion, cardiac involvement in mitochondrial encephalomyopathies varies between 10 and 50 %, when we select patients from different reports (Tab.1).

7.8 Diagnosis of mitochondrial cardiomyopathies

In all cases of infantile onset of cardiomyopathy, irrespective of the type of cardiomyopathy, an OXPHOS disturbance should be ruled out. Respiratory chain defects are most often associated with hypertrophic cardiomyopathies [35], Barth syndrome is associated with left ventricular noncompaction [11]. The most direct approach for the diagnosis of

Tab. 1. Frequency of mitochondrial cardiomyopathies

	Patients (n)	Cardiomyopathy	Arrhythmias	Total	
▪Tulinius [115, 116]	25	4	8	25/12	48 %
▪Böhles	[12]	16	3	16/3	19 %
▪Innsbruck	22	2	4	22/6	27 %
▪Literature	143	12	4	143/16	11%

mitochondrial cardiomyopathy is investigation of an endomyocardial biopsy [91]. An optimal diagnostic strategy would be measurement of mitochondrial function in intact heart muscle mitochondria with respirometry followed by measurement of single enzymes. With micromethods such as PCR, immunoassays and immunocytochemistry, all of them requiring only few micrograms of tissue, a further characterization, e.g. at the DNA level, is possible. A number of genetic defects can be detected from blood directly without endomyocardial biopsy. So far, polarographic investigations of intact mitochondria have been possible only in autopsy material since quite a high amount of tissue is necessary for isolation of mitochondria. Permeabilization of heart muscle fibres [120] in combination with high resolution respirometry [37] allows functional characterization of heart muscle mitochondria in a relatively small tissue sample (5-10 mg) [56, 120]. This method in combination with DNA analysis presents the future possibilities to further characterize primary and secondary mitochondrial defects.

In many cases of mitochondrial cardiomyopathy, skeletal muscle involvement or systemic involvement can be observed. The most important laboratory parameter is lactic acid in different body fluids. It is crucial to strictly observe the sampling conditions. Determination of the lactate/pyruvate ratio is useful when the cytoplasmic redox status is judged. Determination of the β-OH-butyric acid/acetoacetic acid ratio is an oportunity to obtain information about the intramitochondrial redox status. In summary the following investigations should be performed when mitochondrial cardiomyopathy is suspected [3, 84, 124]:

1. Measurement of lactate, pyruvate, β-OH-butyric acid, acetoacetic acid and carnitine in plasma, organic acids in urine, etc. [3, 124]. In some patients, exercise and loading tests can be helpful.

2. ECG, echography of the heart, if possible: ^{31}P-spectroscopy

3. Endomyocardial biopsy (skeletal muscle or other organ biopsies in case of systemic involvement):
   - Histology, histochemistry, immunocytochemistry, electron microsocopy
   - Respirometry in isolated mitochondria or permeabilized muscle fibers or measurement of substrate oxidation in isolated mitochondria or supernatant (600 g).
   - Enzyme determination
   - SDS-PAGE and immunoblotting (subunit pattern of OXPHOS enzymes)
   - DNA diagnostics (mitochondrial and nuclear DNA analysis) (also from blood cells)

It is important to exclude cardiac involvement in every case of encephalomyopathy, and vice versa it seems reasonable to thoroughly investigate skeletal muscle in every case of cardiomyopathy [20, 62, 74, 125]. In many cases, both cardiac and skeletal muscle is affected [44, 46, 62, 69, 75, 114, 131] but cardiac symptoms rarely predominate [20]. Because it has been difficult to obtain sufficient heart muscle for extended diagnostics, it has been attempted in cases of cardiomyopathy to obtain diagnosis by analysis of skeletal muscle [125]. ^{31}P-spectroscopy of the heart gives valuable information [21] but it is not yet routinely available.

In the case of Barth syndrome (BTHS), excretion of 3-methylglutaconic acid, G4.5 gene analysis and cardiolipin measurement [117] can be performed.

7.9 Therapy

Therapy of mitochondrial defects is very difficult. There are only individual reports about a successful therapy of encephalomyopathies [31]. There is the possibility of cofactor substitution. Mainly the substitution of coenzyme Q10 [33, 84], idebenone [59], riboflavin, thiamine, biotin and lipoic acid may be discussed [87]. Also menadione and vitamin C can be used and L-carnitine is considered useful for the prevention of secondary carnitine insufficiency. Dichloracetate (DCA) successfully lowers the plasma lactic acid concentration but there is only limited clinical success and may lead to serious neurological sideeffects during long-term use. Only recently has some beneficial effect of DCA on myocardial dysfunction been described which has been atributed to the inhibiting action of DCA on PDH kinase and in consequence to an increased pyruvate oxidation rate [8]. In Barth syndrome, one single report on successful therapy with pantothenic acid exists [81].

In patients with Kearns-Sayre syndrome, the use of a heart pacemaker has been established. Even prophylactic pacemaker treatment has been recommended for these patients [88].

A patient with Kearns-Sayre syndrome and rapid development of congestive heart failure required heart transplantation [18]. Recently the successful heart transplantation was reported in two patients with mitochondrial myopathies and severe cardiomyopathy in a final state [17].

7.10 Summary and outlook

Many components are able to influence the cardiac OXPHOS system: natural aging, secondary disturbances caused by radical formation after extreme physical activity, malnutrition, hypoperfusion, hypoxia and finally genetic disturbances. Only the scrupulous investigation of myocardial dysfunction from different points of view will allow a deeper knowledge of the role of OXPHOS system disturbances in cardiac disease. At present there is an enormous and confusing amount of individual data at the mitochondrial level in the field of cardiomyopathies. Abnormal mitochondria are not always associated with a biochemical defect. Equally a gene defect may lack a detectable deficiency at the enzymatic level. The same respiratory chain defect may or may not be be associated with cardiomyopathy (i.e. COX deficiency). On the other hand hypertrophic cardiomyopathy may present with different mutations and enzymatic defects. Thus it is impossible to correlate genotype and phenotype. It is important to emphasize that DNA investigations alone cannot explain the cause of cardiomyopathies. It is indispensable to investigate the function of the heart mitochondria. Exact combined characterizations of the OXPHOS system in cardiomyopathies at the functional as well as at the DNA level will be necessary to obtain a deeper insight into the pathophysiology of this heterogeneous group of disorders.

Acknowledgements

I am grateful to Ron de Nijs for frequency studies and Inge Patauner for editorial assistance. The author is supported by grants from the Austrian Research foundation (P 8293 MED) and Milupa International.

List of abbreviations:

OXPHOS – oxidative phosphorylation
PCR – polymerase chain reaction
CPEO – chronic progressive external ophthalmoplegia
MELAS – mitochondrial encephalomyopathy, lactic acidosis, stroke like episodes,
MERRF – myoclonus epilepsy, ragged red fibres
LIMD – lethal infantile mitochondrial disease
LHON – Leber's hereditary neuropathy of the optic nerve
COX – cytochrome c oxidase
DCA – dichloroacetate
WPW – Wolff-Parkinson-White
LGL – Lown-Ganong-Levine
SDS-PAGE – sodium dodecyl sulphate-polyacrylamide gel electrophoresis

References

[1] Adachi, K., Y. Fujiura, F. Myaumi, A. Nozuhara, Y. Sugiu, T. Sakanashi, T. Hidaka, H. Toshima: A deletion of mitochondrial DNA in murine doxorubicin-induced cardiotoxicity. Bioch. Biophys. Res. Comm. **195**, 945-951 (1993).

[2] Ades, L. C., A. K. Gedeon, M. J. Wilson, M. Latham, M. W. Partington, J. C. Mulley, J. Nelson, K. Lui, D. O. Sillence: Barth syndrome: clinical features and confirmation of gene location to distal Xp 28. Am. J. Med. Genet. **45**, 327-334 (1993).

[3] Anan, R., M. Nakagawa, I. Higuchi, S. Nakao, K. Nomoto, H. Tanaka: Deletion of mitochondrial DNA in the endomyocardial biopsy sample from a patient with Kearns-Sayre syndrome. Europ. Heart. J. **13**, 1718-1719 (1992).

[4] Angelini, C., G. F. Micaglio, P. Sforza, P. Melacini, R. Carrozzo, M. Fanin, A. Ferrarese, M. Rosa: Familial lipid storage cardiomyopathy with mitochondrial complex II defect. Neurology **38**, 152 (1988).

[5] Asano, K., M. Nakamura, T. Sato, H. Tauchi, A. Asano: Age dependency of mitochondrial DNA decrease differs in different tissues of rat. J. Biochem. **114**, 303-306 (1993).

[6] Barth, P. G., H. R. Scholte, J. A. Berden, J. M. Van der Klei-Van Moorsel, I. E. M. Luyt-Houwen, E. Th. Van T Veer-Korthof, J. J. Van der Harten, M. A. Sobotka-Plojhar: An X-linked mitochondrial disease affecting cardiac muscle, skeletal muscle and neutrophil leucocytes. J. Neurol. Sci. **62**, 327-355 (1983).

[7] Berko, B. A., M. Swift: X-linked dilated cardiomyopathy. N. Engl. J. Med. **316**, 1186-1191 (1987).

[8] Bersin, R.M., C. Wolfe, M. Kwasman, D. Lau, C. Klinski, K. Tanaka, P. Khorrami, G.N. Henderson, T. De Marco, K. Chatterjee: Improved hemodynamic function and mechanical efficiency in congestive heart failure with sodium dichloroacetate. J. Am. Coll. Cardiol. **23**, 1617-1624 (1994).

[9] Bet, L., N. Bresolin, M. Moggio, G. Meola, S. Jann, F. Rubboli, L. Geremia, G. Scarlato: Biochemical, histochemical, and tissue culture studies in a case of mitochondrial myopathy due to complex I deficiency. Neurology **38**, 188, PP169 (1988).

[10] Bione, S, P. D'Adamo, E. Maestrini, A. K. Gedeon, P. A. Bolhuis, D. Toniolo: A novel X-linked gene, G4. 5 (sic), is responsible for Barth syndrome. Nature. Genet. **12**, 385-389 (1996).

[11] Bleyl, S. B., B. R. Mumford, M.C. Brown-Harrison, L. T. Pagotto, J. C. Carey, T. J. Pysher, K. Ward, T. K. Chin: Xq28-linked noncompaction of the left ventricular myocardium: prenatal diagnosis and pathologic analysis of affected individuals. Am. J. Med. Genet. **72**, 257-265 (1997).

[12] Böhles, H., H. Singer, W. Ruitenbeek, J. M. F. Trijbels, R. C. A. Sengers, R. P. Ketelsen, E. Wagner-Thiessen, H. Wick: Foamy myocardial transformation in a child with a disturbed respiratory chain. Eur. J. Pediatr. **146**, 582-586 (1987).

[13] Bolhuis, P. A., P. G. Barth, F. A. Wijburg, K. M. C. Sinjorgo, W. Ruitenbeek: Molecular basis of mitochondrial myopathies. Lancet **16**, 884 (1988).

[14] Bolhuis, P. A., G. W. Hensels, T. J. M. Hulsebos, F. Baas, P. G. Barth: Mapping of the locus for X-linked cardioskeletal myopathy with neutropenia and abnormal mitochondria (Barth syndrome) to Xq28. Am. J. Hum. Genet. **48**, 481-485 (1991).

[15] Breningstal, G. N.: Approach to diagnosis of oxidative metabolism disorders. Pediatr. Neurol. **9**, 81-90 (1993).

[16] Buchwald, A., H. Till, C. Unterberger, R. Oberschmidt, H. R. Figulla, V. Wiegand: Alterations of the mitochondrial respiratory chain in human dilated cardiomyopathy. Eur. Heart. J. **11**, 509-516 (1990).

[17] Bussieres, L. M., P. W. Pflugfelder, C. Guiraudon, W. F. Brown, D. G. Munoz, A. W. Taylor, W. J. Kostuk: Exercise responses after cardiac transplantation in mitochondrial myopathy. Am. J. Card. **71**, 1003-1006 (1993).

[18] Channer, K. S., J. L. Channer, M. J. Campbell, R. J. Russel: Cardiomyopathy in the Kearns-Sayre syndrome. Br. Heart. J. **59**, 486-90 (1988).

[19] Colan, S. D., P.J. Spevak, I. A. Parness, A. S. Nadas: Cardiomyopathies. In: Nadas' Pediatric Cardiology, D. C. Fyler (ed.). Hanley &Belfus Inc., Philadelphia, pp. 329-361. (1992).

[20] Constans, J., A. LeHerissier, M. Coquet, J. P. Mazat, T. Letellier, P. Durandet, R. Roudaut, P. Gosse, C. Conri, M. Dallocchio: Ventricular arrhythmia revealing

mitochondrial myopathy in a 69-year-old woman. Europ. Heart. J. **14**, 1137-1139 (1993).
[21] Conway, M. A., J. Allis, R. Ouwerkerk, T. Niioka, B. Rajagopalan, G. K. Radda: Detection of low phosphocreatine to ATP ratio in failing hypertrophied human myocardium by 31P magnetic resonance spectroscopy. Lancet **338**, 973-976 (1991).
[22] Corral, M., Debrinski, G. Stepien, J. M. Shoffner, M. T. Lott, K. Kanter, D. C. Wallace: Hypoxemia is associated with mitochondrial DNA damage and gene induction. Jama **266**, 1812-1816 (1991).
[23] Corral-Debrinski, M., J. M. Shoffner, M. T. Lott, D. C. Wallace: Association of mitochondrial DNA damage with aging and coronary atherosclerotic heart disease. Mut. Res. **275**, 169-180 (1992).
[24] Coster, R. V., A. Lombes, D. C. De Vivo, T. L. Chi, W. E. Dodson, St. Rothman, E. J. Orrechio, W. Grover, G. T. Berry, J. F. Schwartz, A. Habib, S. Di Mauro: Cytochrome c oxidase-associated Leigh syndrome: phenotypic features and pathogenetic speculations. J. Neurol. Sci. **104**, 97-111 (1991).
[25] Adamo, P. D., L. Fassone, A. Gedeon, E. A. Janssen, S. Bione, P. A. Bolhuis, P. G. Barth, M. Wilson, E. Haan, K. H. Orstavik, M. A. Patton, A. J. Green, E. Zammarchi, M. A. Donati, D. Toniolo: The X-linked gene G4.5 is responsible for different infantile dilated cardiomyopathies. Am. J. Hum. Genet. **61**, 862-867 (1997).
[26] Das, A. M., D. A. Harris: Regulation of the mitochondrial ATP synthase is defective in rat heart during alcohol-induced cardiomyopathy. Bioch Bioph Acta 1993, 1181: 295-299.
[27] De Vivo, D. C.: The expanding clinical spectrum of mitochondrial diseases. Brain & Development **15**, 1-22 (1993).
[28] Di Lisa, F., Chong-Zhu Fan, G. Gambassi, B. A. Hogue, I. Kudryashova, R.G. Hansford: Altered pyruvate dehydrogenase control and mitochondrial free Ca2+ in hearts of cardiomyopathic hamsters. Am. J. Physiol. **264**, H2188-2197 (1993).
[29] DiMauro, S., E. Bonilla, A. Lombes, S. Shanske, C. Minetti, C. T. Moraes: Mitochondrial encephalomyopathies. Pediatr. Neurol. **8**, 483-506 (1990).
[30]. Duboc, D., P. Abastado, M. Muffat-Joly, P. Perrier, M. Toussaint, C. Marsac, D. Francois, T. Lavergne, J.-J. Pocidalo, F. Guerin, A. Carpentier: Evidence of mitochondrial impairment during cardiac allograft rejection. Transplantation **59**, 751-755 (1990).
[31] Ferrans, V. J., H. A. McAllister, W. H. Haese: Infantile cardiomyopathy with histiocytoid change in cardiac muscle cells. Circulation **53**, 708-719 (1976).
[32] Figarella-Branger, D., J. F. Pellissier, C. Scheiner, F. Wernert, C. Desnuelle: Defects of the mitochondrial respiratory chain complexes in three pediatric cases with hypotonia and cardiac involvement. J. Neurol. Sci. **108**, 105-113 (1992).
[33] Folkers, K.: Heart failure is a dominant deficiency of coenzyme Q10 and challenges for future clinical research on CoQ10. Clin. Invest. **71**, 51-54 (1993).
[34] Gerbitz, K.-D., B. Obermaier-Kusser, S. Zierz, D. Pongratz, J. Müller-Höcker, P. Lestienne: Mitochondrial myopathies: divergences of genetic deletions, biochemical defects and the clinical syndromes. J. Neurol. **237**, 5-10 (1990).

[35] Guenthard, J., F. Wyler, B. Fowler, R. Baumgartner: Cardiomyopathy in respiratory chain disorders. Arch. Dis. Child. **72**, 223-226 (1995).
[36] Guerrieri, F., G. Capozza, A. Fratello, F. Zanotti, S. Papa: Functional and molecular changes in FoF1 ATP-synthase of cardiac muscle during aging. Cardioscience **4**, 93-98 (1993).
[37] Haller, T., M. Ortner, E. Gnaiger: A respirometer for investigating oxidative cell metabolism: toward optimization of respiratory studies. Anal. Biochem. **218**, 338-342 (1994).
[38] Hattori, K., T. Ogawa, T. Kondo, M. Mochizuki, M. Tanaka, S. Sugiyama, T. Ito, T. Satake, T. Ozawa: Cardiomyopathy with mitochondrial DNA mutations. Am. Heart. J. **122**, 866-869 (1991).
[39] Hattori, K., M. Tanaka, S. Sugiyama, T. Obayashi, T. Ito, T. Satake, Y. Hanaki, J. Asai, M. Nagano, T. Ozawa: Age-dependent increase in deleted mitochondrial DNA in the human heart: Possible contributory factor to presbycardia. Am. Heart. J. **121**, 1735-1742 (1991).
[40] Hayakawa, M., S. Sugiyama, K. Hattori, M. Takasawa, T. Ozawa: Age-associated damage in mitochondrial DNA in human hearts. Mol. Cell. Biochem. **119**, 95-103 (1993).
[41] Hodgson, S., A. Child, M. Dyson: Endocardial fibroelastosis: possible X-linked inheritance. J. Med. Genet. **24**, 210-214 (1987).
[42] Holme, E., J. Greter, C. E. Jacobson, N. G. Larsson, S. Lindstedt, K. O. Nilsson, A. Oldfors, M. Tulinius: Mitochondrial ATP-synthase deficiency in a child with 3-methylglutaconic aciduria. Pediatr. Res. **32**, 731-735 (1992).
[43] Holt, I. J., A. E. Harding, J. A. Morgan-Hughes: Deletions of muscle mitochondrial DNA in patients with mitochondrial myopathies. Nature **331**, 717 (1988).
[44] Hübner, G., R. Grantzow: Mitochondrial cardiomyopathy with involvement of skeletal muscles. Virchows Arch. (Pathol. Anat.) **399**, 115-125 (1993).
[45] Hurko, O., D. R. Johns, S. L. Rutledge, O. C. Stine, P. L. Peterson, N. R. Miller, M. E. Martens, D. B. Drachman, R. H. Brown, C. P. Lee: Heteroplasmy in chronic external ophthalmoplegia: Clinical and molecular observations. Pediatr. Res. **28**, 542-548 (1990).
[46] Ibel, H., W. Endres, H. B. Hadorn, T. Deufel, I. Paetzke, M. Duran, N. G. Kennaway, K. M. Gibson: Multiple respiratory chain abnormalities associated with hypertrophic cardiomyopathy and 3-methylglutaconic aciduria. Eur. J. Pediatr. **152**, 665-670 (1993).
[47] Ino, T., W. G. Sherwood, E. Cutz, L. N. Benson, V. Rose, R. M. Freedom: Dilated cardiomyopathy with neutropenia, short stature, and abnormal carnitine metabolism. J. Pediatr. **11**, 511-514 (1988).
[48] Inui, K., H. Fukushima, T. Tsukamoto, M. Taniike, M. Midorikawa, J. Tanaka, T. Nishigaki, S. Okada: Mitochondrial encephalomyopathies with the mutation of the mitochondrial tRNA Leu(UUR)gene. J. Pediatr. **120**, 62-66 (1992).
[49] Ito, T., K. Hattori, T. Obayashi, M. Tanaka, S. Sugiyama, T. Ozawa: Mitochondrial DNA mutations in cardiomyopathy. Jpn. Circ. J. **56**, 1045-1053 (1992).
[50] Jarcho, J. A., W. McKenna, J. A. P. Pare, S. D. Salomon, R. F. Holcombe, S. Dickie, T. Levi, H. Donis-Kel-

ler, J. G. Seidman, C. E. Seidman: Mapping a gene for familial hypertrophic cardiomyopathy to chromosome 14q1. N. Engl. J. Med. **321**, 1372-1378 (1989).

[51] Jordens, E. Z., L. Palmieri, M. Huizing, L. P. van den Heuvel, R. C. Sengers, A. Dorner, W. Ruitenbeek, F. J. Trijbels, J. Valsson, G. Sigfusson, F. Palmieri, J. A. Smeitink: Adenine nucleotide translocator 1 deficiency associated with Sengers syndrome. Ann. Neurol. **52**, 95-99 (2002).

[52] Kelley, R. I., B. J. Clark, D. H. Morton, W. G. Sherwood: X-linked cardiomyopathy, neutropenia and increased urinary levels of 3-methylglutaconic and 2-ethylhydracrylic acids. Am. J. Hum. Genet. **45**, A7 (1989).

[53] Kelly, D. P., A. W. Strauss: Inherited cardiomyopathies. N Engl J Med 1994, **330**, 913-9.

[54] Kleber, F. X., J.-W. Park, G. Hübner, A. Johannes, D. Pongratz, E. König: Congestive heart failure due to mitochondrial cardiomyopathy in Kearns-Sayre syndrome. Klin. Wochenschr. **65**, 480-486 (1987).

[55] Kohlschütter, A., G. Hausdorf: Primary (genetic) cardiomyopathies in infancy. A survey of possible disorders and guidelines for diagnosis. Eur. J. Pediatr. **145**, 454-459 (1986).

[56] Kunz, W. S., A. V. Kuznetsov, W. Schulze, K. Eichhorn, L. Schild, F. Striggow, R. Bohnensack, S. Neuhof, H. Grasshoff, H. W. Neumann, F. N. Gellerich: Functional characterization of mitochondrial oxidative phosphorylation in saponin-skinned human muscle fibers. Biochim. Biophys. Acta. **1144**, 46-53 (1993).

[57] Lamperth, L., M. C. Dalakas, F. Dagani, J. Anderson, R. Ferrari: Abnormal skeletal and cardiac muscle mitochondria induced by zidovudine (AZT) in human muscle in vitro and in an animal model. Laboratory Investigation **65**, 742-751 (1991).

[58] Lees, M. H.: Perinatal asphyxia and the myocardium. J Pediatr **96**, 675-678 (1980).

[59] Lerman-Sagie, T., P. Rustin, D. Lev, M. Yanoov, E. Leshinsky-Silver, A. Sagie, T. Ben-Gal, A. Munnich: Dramatic improvement in mitochondrial cardiomyopathy following treatment with idebenone. J. Inherit. Metab. Dis. **24**, 28-34 (2001).

[60] Lewis, W., B. Gonzalez, A. Chomyn, T. Papoian: Zidovudine induces molecular, biochemical, and ultrastructural changes in rat skeletal muscle mitochondria. J. Clin. Invest. **89**, 1354-1360 (1992).

[61] Linnane, A. W., S. Marzuki, T. Ozawa, M. Tanaka: Mitochondrial DNA mutations as an important contributor to aging and degenerative diseases. Lancet, 642-645 (1989).

[62] Mackay, E. H., R. S. Brown, D. Pickering: Cardiac biopsy in skeletal myopathy: report of a case with myocardial mitochondrial abnormalities. J. Path. **120**, 35-42 (1976).

[63] Marin-Garcia, J., M.J.Goldenthal: Cardiomyopathy and abnormal mitochondrial function. Cardiovascular Research **28**, 456-463 (1994).

[64] Marin-Garcia, J., M. J. Goldenthal: Mitochondrial cardiomyopathy: Molecular and biochemical analysis. Pediatr. Cardiol. **18**, 251-260 (1997).

[65] Mastaglia, F. L., P. L. Thompson, J. M. Papadimitriou: Mitochondrial myopathy with cardiomyopathy, lactic acidosis and responses to prednisone and thiamine. Aust. N Z J Med. **10**, 660-664 (1980).

[66] Maurer, I., S. Zierz: Myocardial respiratory chain enzyme activities in idiopathic dilated cardiomyopathy and comparison with those in atherosclerotic coronary artery disesase and valvular aortic stenosis. Am. J. Cardiol. **72**, 428-433 (1993).

[67] Maurer I., S. Zierz: Mitochondrial respiratory chain enzyme activities in tetralogy of Fallot. Clinical Investigator **72**, 358-363 (1994).

[68] Moraes, C. T., E. A. Schon, S. DiMauro. Mitochondrial diseases: toward a rational classification. In: Current Neurology. Mosby Year Book, Appal, S. (ed.). St. Louis: pp 83-119 (1991).

[69] Moreadith, R. W., M. L. Batshaw, T. Ohnishi, D. Kerr, B. Knox, D. Jackson, R. Hruban, J. Olson, B. Rynafarje, A. L. Lehninger: Deficiency of the iron-sulfur clusters of mitochondrial reduced nicotinamide-adenine dinucleotide-ubiquinone oxidoreductase (complex I) in an infant with congenital lactic acidosis. J. Clin. Invest. **74**, 685-697 (1984).

[70] Müller-Höcker, J.: Cytochrome-c-oxidase deficient cardiomyocytes in the human heart. An age related phenomenon. Am. J. Path. **134**, 1167-1173 (1989).

[71] Müller-Höcker, J., H. Ibel, I. Paetzke, T. Deufel, W. Endres, B. Kadenbach, J. M. Gokel, G. Hübner: Fatal infantile mitochondrial cardiomyopathy and myopathy with heterogeneous tissue expression of combined respiratory chain deficiencies. Virchows. Arch. Pathol. Anat. **419**, 355-362 (1991).

[72] Müller-Höcker, J., P. Seibel, K. Schneiderbanger, Ch. Zietz, B. Obermaier-Kusser, K. D. Gerbitz, B. Kadenbach: In situ hybridization of mitochondrial DNA in the heart of a patient with Kearns-Sayre syndrome and dilatative cardiomyopathy. Hum. Pathol. **23**, 1431-1437 (1992).

[73] Müller- Höcker, J., G. Hübner, K. Bise, Ch. Förster, St. Hauck, I. Paetzke, D. Pongratz, B. Kadenbach: Generalized mitochondrial microangiopathy and vascular cytochrome c oxidase deficiency. Arch. Pathol. Lab. Med. **117**, 202-210 (1993).

[74] Munnich, A., P. Rustin, A. Rötig, D. Chretien, J.-P. Bonnefont, C. Nuttin, V. Cormier, A. Vassault, P. Parvy, J. Bardet, C. Charpentier, D. Rabier, J.-M. Saudubray: Clinical aspects of mitochondrial disorders. J. Inher. Metab. Dis. **15**, 448-455 (1992).

[75] Nagai, T., Y. Tuchiya, Y. Taguchi, R. Sakuta, T. Ichiki, I. Nonaka: Fatal infantile mitochondrial encephalomyopathy with complex I and IV deficiencies. Pediatr. Neurol. **9**, 151-154 (1993).

[76] Neustein, H. B., P. R. Lurie, B. Dahms, M. Takahashi: An X-linked recessive cardiomyopathy with abnormal mitochondria. Pediatrics **64**, 24-29 (1979).

[77] Nikoskelainen, E., O. Wanne, M. Dahl: Pre-excitation syndrome and Leber's hereditary optic neuroretinopathy. Lancet **23**, 696 (1985).

[78] Nikoskelainen, E. K., M.-L. Savontaus, O. P. Wanne, M. J. Katila, K. U. Nummelin: Leber's hereditary optic neuroretinopathy, a maternally inherited disease. Arch. Ophthalmol. **105**, 665-671 (1987).

[79] Obayashi, T., K. Hattori, S. Sugiyama, M. Tanaka, T. Tanaka, S. Itoyama, H. Deguchi, K. Kawamura, Y. Koga, H. Toshima, N. Takeda, M. Nagano, T. Ito, T. Ozawa: Point mutations in mitochondrial DNA in patients with hypertrophic cardiomyopathy. Am. Heart. J. **124**, 1263-1269 (1992).

[80] Oldfors, A., H. Sommerland, E. Holme, M. Tulinius, B. Kristiansson: Cytochrome c oxidase deficiency in infancy. Acta. Neuropathol. **77**, 267-275 (1989).
[81] Ostman-Smith, I., G. Brown, A. Johnson, J. M. Land: Dilated cardiomyopathy due to type II X-linked 3-methylglutaconic aciduria: successful treatment with pantothenic acid. Brit. Heart J. **72**, 349-353 (1994).
[82] Örstavik, K. H., F. Skjörten, M. Hellebostad, P. Haga, A. Langslet: Possible X-linked congenital mitochondrial cardiomyopathy in three families. J. Med. Genet. **30**, 269-272 (1993).
[83] Ozawa, T., M. Tanaka, S. Sugiyama, K. Hattori, T. Ito, K. Ohno, A. Takahashi, W. Sato, G. Takada, B. Mayumi, K. Yamamoto, K. Adachi, Y. Koga, H. Toshima: Multiple mitochondrial DNA deletions exist in cardiomyocytes of patients with hypertrophic or dilated cardiomyopathy. Bioch. Biophys. Res. Commun. **170**, 830-836 (1990).
[84] Ozawa, T.: Mitochondrial cardiomyopathy. Herz **19**, 105-118 (1994).
[85] Papadimitriou, A., H. B. Neustein, S. DiMauro, R. Stanton, N. Bresolin: Histiocytoid cardiomyopathy of infancy: deficiency of reducible cytochrome b in heart mitochondria. Pediatr. Res. **18**, 1023-1028 (1984).
[86] Pastores, G. M., F. M. Santorelli, S. Shanske, B. D. Gelb, B. Fyfe, D. Wolfe, J. P. Willner: Leigh syndrome and hypertrophic cardiomyopathy in an infant with a mitochondrial DNA point mutation (T89993G). Am. J. Med. Genetics **50**, 265-271 (1994).
[87] Przyrembel, H.: Therapy of mitochondrial disorders. J. Inher. Metab. Dis. **10**, 129-146 (1987).
[88] Remes, A. M., I. E. Hassinen, K. Majamaa, K. J. Peuhkurinen: Mitochondrial DNA deletion diagnosed by analysis of an endomyocardial biopsy specimen from a patient with Kearns-Sayre syndrome and complete heart block. Br Heart. J. **68**, 408-411 (1992).
[89] Rimoldi, M., E. Bottachi, L. Rossi, F. Cornelio, G. Uziel, S. DiDonato: Cytochrome c oxidase deficiency in muscles of a floppy infant without mitochondrial myopathy. J. Neurol. **227**, 201-207 (1982).
[90] Romero, N. B., C. Marsac, M. Paturneau-Jouas, H.S. Ogier, M. Magnier Fardeau: Infantile familial cardiomyopathy due to mitochondrial complex I and IV associated deficiency. Neuromusc. Disord. **3**, 31-42 (1993).
[91] Rustin, P., J. Lebidois, D. Chretien, T. Bourgeron, J. F. Piechaud, A. Rötig, A. Munnich, D. Sidi: Endomyocardial biopsies for early detection of mitochondrial disorders in hypertrophic cardiomyopathies. J. Pediatr. **124**, 224-228 (1994).
[92] Rutledge, J. C., J. E. Haas, R. Monnat, J. M. Milstein: Hypertrophic cardiomyopathy is a component of subacute necrotizing encephalomyelopathy. J. Pediatr. **101**, 706-710 (1982).
[93] Sengers, R. C. A., B. G. A. ter Haar, J. M. F. Trijbels, J. L. Willems, O. Daniels, A. M. Stadhouders: Congenital cataract and mitochondrial myopathy of skeletal and heart muscle associated with lactic acidosis after exercise. J. Pediatr. **86**, 873-880 (1975).
[94] Sengers, R. C. A., J. M. F. Trijbels, A. J. M. Bakkeren, W. Ruitenbeek, J. C. Fischer, A. J. M. Janssen, M. Stadhouders, H. J. Laak: Deficiency of cytochromes b and a3 in muscle from a floppy infant with cytochrome

oxidase deficiency. Eur. J. Pediatr. **141**, 178-180 (1984).
[95] Sengers, R. C .A., A.M. Stadhouders, E. Van Lakwijk-Vondrovicova, K. Kubat, W. Ruitenbeek: Hypertrophic cardiomyopathy associated with mitochondrial myopathy of voluntary muscles and congenital cataract. Br. Heart. J. **54**, 543-547 (1985).
[96] Shoffner, J., M. Lott, A. Lezza, P. Seibel, S. Ballinger, D. C. Wallace. Myoclonic epilepsy and ragged-red fiber disease (MERRF) is associated with a mitochondrial DNA tRNALYS mutation. Cell **61**, 931-937 (1990).
[97] Shoffner, J. M., D. C. Wallace: Oxidative phosphorylation diseases. Disorders of two genomes. Adv. Human. Genetics. **19**, 267-330 (1990).
[98] Silver, M. M., J. E. Burns, R. K. Sethi, R. D. Rowe: Oncocytic cardiomyopathy in an infant with oncocytosis in exocrine and endocrine glands. Hum. Pathol. **11**, 598-605 (1980).
[99] Silvestri, G., F. M. Santorelli, S. Shanske, C. B. Whitley, L. A. Schimmenti, S. A. Smith, S. DiMauro: A new mtDNA mutation in the tRNALeu(uur) gene associated with maternally inherited cardiopathy. Hum. Mut. **3**, 37-43 (1994).
[100] Silvestri, G., E. Ciafaloni, F. M. Santorelli, S. Shanske, S. Servidei, W. D. Graf, M. Sumi, S. DiMauro: Clinical features associated with the A-G transition at nucleotide 8344 of mtDNA ("MERRF point mutation"). Neurology **43**, 1200-1206 (1993).
[101] Singal, P. K., C. M. R. Deally, L. E. Weinberg: Subcellular effects of adriamycin in the heart: a concise review. J. Mol. Cell. Cardiol. **19**, 817-882 (1987).
[102] Smeitink, J. A. M., R. C. A. Sengers, J. M. F. Trijbels, W. Ruitenbeek, O. Daniels, A. M. Stadhouders, M. J. H. Kock-Jansen: Fatal neonatal cardiomyopathy associated with cataract and mitochondrial myopathy. Eur. J. Pediatr. **148**, 656-659 (1989).
[103] Sperl, W., W. Ruitenbeek, J. M. F Trijbels, G. C. Korenke, R. C. A. Sengers: Heterogeneous tissue expression of enzyme defects in mitochondrial myopathies. J. Inher. Metab. Dis. **13**, 359-362 (1990).
[104] Sperl, W., D. Skladal, N. Lanznaster, R. Schranzhofer, G. Zaunschirm, E. Gnaiger, F. Gellerich: Polarographic studies of saponin-skinned muscle fibres in patients with mitochondrial myopathies. J. Inher. Metab. Dis. **17**, 307-310 (1994).
[105] Stumpf, D. A.: Friedreich's disease: a metabolic cardiomyopathy. Am. Heart. J. **104**, 887-888 (1982).
[106] Suomalainen, A., A. Paetau, H. Leinonen, A. Majander, L. Peltonen, H. Somer: Inherited idiopathic dilated cardiomyopathy with multiple deletions of mitochondrial DNA. Lancet **340**, 1319-20 (1992).
[107] Sweetman, L., J. C. Williams: Branched chain organic acidurias. In: The metabolic and molecular bases of inherited disease. Scriver, Beaudet, Valle, Sly, Childs, Kinzler, Vogelstein (eds.). McGraw-Hill Companies, 8th ed., pp. 2125-2163.
[108] Takeda, N., A. Tanamura, T. Iwai, I. Nakamura, M. Kato, T. Ohkubo, K. Noma: Mitochondrial DNA deletion in human myocardium. Molecular and Cellular Biochemistry **119**, 105-108 (1993).
[109] Tanaka. M., M. Nishikimi, H. Suzuki, T. Ozhawa, M. Nishizawa, K. Tanaka, T. Miyatake: Deficiency of subunits in heart mitochondrial NADH-ubiquinone oxi-

doreductase of a patient with mitochondrial encephalomyopathy and cardiomyopathy. Biochem. Biophys. Res. **140**, 88-93 (1986).
[110] Tanaka, K., T. Yoshimura, H. Muratani, J. Kira, Y. Itoyama, I. Goto: Familial myopathy with scapulohumeral distribution, rigid spine, cardiopathy and mitochondrial abnormality. J. Neurol. **236**, 52-54 (1989).
[111] Tanaka, M., H. Ino, K. Ohno, K. Hattori, W. Sato, T. Ozawa, T. Tanaka, S. Itoyama: Mitochondrial mutation in fatal mitochondrial cardiomyopathy. Lancet ii, **336**, 1452 (1990).
[112] Taniike, M., H. Fukushima, I. Yanagihara, H. Tsukamoto, J. Tanaka, H. Fujimura, T. Nagai, T. Sano, K. Yamaoka, K. Inui, S. Okada: Mitochondrial tRNAIle mutation in fatal cardiomyopathy. Bioch. Biophys. Res. Commun. **186**, 47-53 (1992).
[113] Trijbels, J. M. F., H. R. Scholte, W. Ruitenbeek, R. C. A. Sengers, A. J. M. Janssen, H. F. M. Busch: Problems with the biochemical diagnosis in mitochondrial (encephalo-) myopathies. Eur. J. Pediatr. **152**, 178-184 (1993).
[114] Tulinius, M. H., B. O. Eriksson, O. Hjalmarson, E. Holme, A. Oldfors: Mitochondrial myopathy and cardiomyopathy in siblings. Pediatr. Neurol. **5**, 182-8 (1989).
[115] Tulinius, M. H., E. Holme, B. Kristiansson, N.-G Larsson, A. Oldfors: Mitochondrial encephalomyopathies in childhood. I. Biochemical and morphologic investigations. J. Pediatr. **119**, 242-250 (1991).
[116] Tulinius, M. H., E. Holme, B. Kristiansson, N.-G Larsson, A. Oldfors: Mitochondrial encephalomyopathies in childhood. II. Clinical manifestation and syndromes. J. Pediatr. **119**, 251-259 (1991).
[117] Valianpour, F., R. J. Wanders, H. Overmars, P. Vreken, A. H. Van Gennip, F. Baas, P. Plecko, R. Santer, K. Becker, P. G. Barth: Cardiolipin deficiency in X-linked cardioskeletal myopathy and neutropenia (Barth syndrome, MIM 302060): a study in cultured skin fibroblasts. J. Pediatr. **141**, 729-733 (2002).
[118] Van Ekeren, G. J., A. M. Stadhouders, G. J. M. Egberink, R. C. A. Sengers, O. Daniels, K. Kubat: Hereditary mitochondrial hypertrophic cardiomyopathy with mitochondrial myopathy of skeletal muscle, congenital cataract and lactic acidosis. Virchows. Arch. A. **412**, 47-52 (1987).
[119] Van Ekeren, G. J., A. M. Stadhouders, J. A. M. Smeitink, R. C. A. Sengers: A retrospective study of patients with the hereditary syndrome of congenital cataract, mitochondrial myopathy of heart and skeletal muscle and lactic acidosis. Eur. J. Pediatr. **152**, 255-259 (1993).
[120] Veksler, V. I., A. V. Kuznetsov, V. G. Sharov et al.: Mitochondrial respiratory parameters in cardiac tissue: a novel method of assessment by using saponin-skinned fibers. Bioch. Biophys. Acta. **892**, 191-196 (1987).
[121] Voth, D.: Über die Arachnozytose des Herzmuskels. Frankfurter Zeitschrift für Pathologie **71**, 646-656 (1962).
[122] Wallace, D. C., et: Mitochondrial DNA mutation associated with Leber's hereditary optic neuropathy. Science **242**, 1427 (1988).
[123] Wallace, D.C.: Diseases of the mitochondrial DNA. Annu. Rev. Biochem. **61**, 1175-1212 (1992).
[124] Wijburg, F. A., Mitochondrial encephalomyopathies: diagnostic and therapeutic studies. Thesis, University of Amsterdam (1993).
[125] Wimmer, M.: Department of Cardiology, Childrens Hospital, University of Vienna, personal communication.
[126] Yang, F. Y., Z. H. Lin, S. G. Li, B. Q. Guo, Y. S. Yin: Keshan Disease - an endemic mitochondrial cardiomyopathy in China. J. Trace. Elem. Electrolytes Health Dis. **2**, 157-163 (1988).
[127] Yoon, K. L., S. G. Ernst, C. Rasmussen, E. C. Dooling, J. R. Aprille: Mitochondrial disorder associated with newborn cardiopulmonary arrest. Pediatr. Res. **33**, 433-440 (1993).
[128] Zenner, K., R. Gold, B. Meurers, H. Reichmann: Die mitochondrialen Enzephalomyopathien. Kearns-Sayre-Syndrom, MELAS und MERRF im Vergleich. Nervenarzt **61**, 597-603 (1990).
[129] Zeviani, M., D. Van Dyke, S. Servidei, S. C. Bauserman, E. Bonilla, E. T. Beaumont, J. Sharda, K. Vanderlaan, S. DiMauro: Myopathy and fatal cardiopathy due to cytochrome c oxidase deficiency. Arch. Neurol. **43**, 1198-1202 (1986).
[130] Zeviani, M., C. Gellera, C. Antozzi, M. Rimoldi, L. Morandi, F. Villani, V. Tiranti, St. Di Donato: Maternally inherited myopathy and cardiomyopathy: association with mutation in mitochondrial DNA tRNALeu(UUR). Lancet **338**, 143-147 (1991).
[131] Zheng, X., J. M. Shoffner, M. T. Lott, A. S. Voljavec, N. S. Krawiecki, K. Winn, D. C. Wallace: Evidence in a lethal infantile mitochondrial disease for a nuclear mutation affecting respiratory complexes I and IV. Neurology **39**, 1203-1209 (1989).

8 Pericardial effusion refractive to therapy in an infant with the carbohydrate-deficient glycoprotein syndrome

W. Kienast, F. Walther, K. Heyne

Non infectious pericardial effusions are extremely rare in infancy and only individual cases have been described in patients with hypothyroidism and glycogen storage disease. However, since 1988 an increasing number of even emergency pericardiocenteses has been reported [1].

We observed an 8-week-old infant with a pericardial effusion, detected during routine sonography of the upper abdomen. The liver was palpable 2 cm below the right costal margin. There was a slight tachycardia of 143 bpm and a decreased blood pressure amplitude (97/73 mm Hg). Heart sounds were weak. X-ray of the thorax showed cardiomegaly (CTI 0.8). Pericardial puncture produced 72 ml of clear fluid with a protein concentration of 32.6 mg/l. Cytological analysis revealed reticulohistiocytes presenting storage phenomena, erythrocytes and vacuolated mesothelial cells negative for PAS and alcian blue stain. There was no bacterial growth.

From the pericard 20 to 55 ml could be drained per day until the 6th day when drainage could be halted, however the pericardial effusion persisted despite equilibration of hypoproteinaemia (39 g/l) and diuretic therapy (Fig. 1). After the intermittent removal of 72 ml of effusion fluid, a pericardial window was operatively performed.

Carbohydrate-deficient glycoprotein syndrome was suspected based an the major symptoms according to Petersen et al. [4] namely: dystrophy, weak sucking reflex, psychomotor retardation, hypotonia, inverted nipples, lipodystrophy, hepatomegaly and cerebellar hypoplasia. Both kidneys were considered enlarged according to sonographic measurements. Aminotransferase activities were increased. The concentrations of the following serum glycoproteins were decreased: haemosiderin (25 gmol/1) and transferrin (0.92 g/l) in connection with anaemia (haematocrit 27%; haemoglobin 5.5 mmol/1). Thyroxine binding globulin (4.6 mg/l) was decreased with normal concentrations of TSH, free T4, free T3 and T3. T4 concentrations were slightly decreased. The decreased apolipoprotein B concentration (0.468 g/l) was accompanied by low concentrations of LDL- (0.55 mmol/ 1) and HDL-cholesterol (0.40-0.56 mmol/l). The concentration of anti-thrombin III was 14% of normal. The diagnosis was confirmed by demonstrating undersialylated variants of the following glycoproteins: α1-antitrypsin, α1-acid glycoprotein (orosomucoid), α2-HS-glycoprotein and transferrin using isoelectric focussing (Prof. Weidinger; Munich).

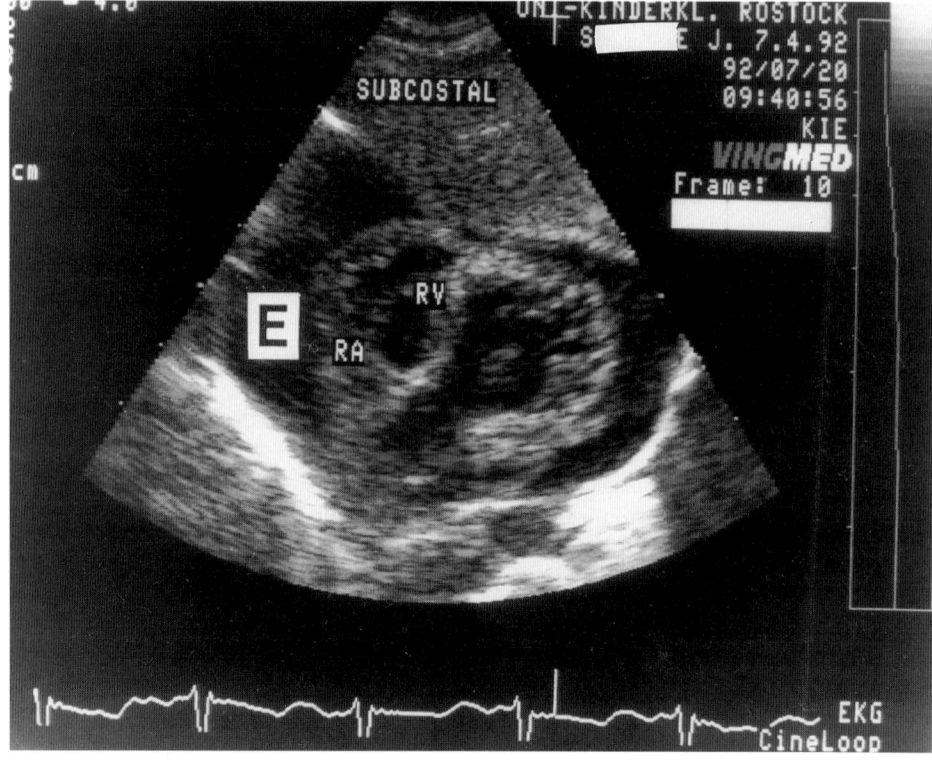

Fig. 1: Persisting pericardial effusion (E) in a 14-week-old infant with carbohydrate deficient glycoprotein syndrome. RV = right ventricle; RA = right atrium.

The aetiology of this disease is still unclear. Inheritance is considered autosomal heterozygous [4], which is supported by the observations within different families [2]. Recently a probable incomplete autosomal dominant mode has been proposed. Symptoms vary considerably with patient age. Dramatic forms, named "infantile alarming multisystem stage" with 15-20% mortality are characteristic in infancy [3]. Apart from the mentioned pericardial effusion, liver failure, apoplectic attacks and severe infections, in particular pneumonia have been reported. Despite the relative lack of clinical symptoms in our case of pericardial effusion, several casualties following heart tamponade have been reported [1, 3].

References

[1] Harding, B. N., D. B. Dunger, D. B. Grant, M. Erdohazi. Familial olivopontocerebellar atrophy with neonatal onset: a recessive inherited syndrome with systemic and biochemical abnormalities. J. Neurol. Neurosurg. Psychiatry **51**, 385-390 (1988).

[2] Jaeken, J., M. Vanderschueren-Lodewyeck, P. Casaer, L. Snoeck, L. Corbeel, E. Eggermont, R. Eeckels: Familial psychomotor retardation with markedly fluctuating serum prolactin, FSH and GH levels, partial TBG deficiency, increased serum aryl-sulfatase A and increased CSF protein: a new syndrome? Pediatr. Res. **14**, 170 (1980).

[3] Jaeken, J., B. Hagberg, P. Stromme: Clinical presentation and natural course of the carbohydrate-deficient glycoprotein syndrome. Acta Paediatr. Scand. Suppl. **375**, 6-13 (1991).

[4] Petersen, M. B., K. Brostrom, H. Stibler, F. Skovby: Early manifestations of the carbohydrate-deficient glycoprotein syndrome. J. Pediatr. **122**, 66-70 (1993).

9 Cardiomyopathy in congenital disorders of glycosylation (CDG)

J. Gehrmann, H. Böhles, T. Marquardt

Congenital disorders of glycosylation (CDG) are a group of inherited metabolic multisystem disorders characterized by defects in the glycosylation of proteins and lipids. Defective glycosylation can result in structural malformation and malfunction of affected glycoproteins. Cardiac involvement is a frequent symptom of these disorders and may be caused by a hypoglycosylation of structural myocardial proteins. In this article, we present several examples of cardiomyopathy in infants and children and summarize the current knowledge of cardiomyopathy in CDG.

Hypertrophic and dilated cardiomyopathy are both observed in CDG patients. Four of our patients presented with dilated cardiomyopathy and all have so far unidentified molecular defects being distinct from the eleven different CDG types known so far (CDG-x). Two patients with the most common form of CDG, CDG-Ia, had hypertrophic cardiomyopathy. Two patients had a fatal outcome related to cardiac disease, one is listed for cardiac transplantation. Three other patients are alive, but have some residual cardiac dysfunction. In three patients cardiomyopathy was the key finding leading to the diagnosis of CDG.

CDG have to be considered in the evaluation of genetic conditions associated with cardiomyopathy. Cardiac involvement contributes significantly to morbidity, mortality, and probably to sudden cardiac death in young children with this disorder. Patients with this condition should undergo routine cardiac examination. As the phenotype of congenital disorders of glycosylation can be highly variable with cardiomyopathy dominating the clinical phenotype in some patients, the presence of congenital disorders of glycosylation should be considered in every child with cardiomyopathy.

9.1 Introduction

CDG are a family of rapidly growing inherited multisystem disorders [8]. They are caused by defects in the biosynthesis of N- or O-linked oligosaccharide chains attached to glycoproteins or glycolipids. Being necessary for the function of many glycoproteins, truncated or missing oligosaccharide side chains often severely affect the structural integrity and function of the affected biomolecules.

Eleven different CDG types based on different molecular defects have been identified up to now. In addition about

30% of the patients have molecular defects different from those already identified (CDG-x). Although the association of cardiomyopathy with genetic disorders is well-known and appreciated in the paediatric cardiology community [14], CDG are almost never considered by cardiologists in the differential diagnosis of disorders leading to cardiomyopathy. Due to a lack of clinical recognition and underreporting, the incidence of cardiomyopathy in these disorders is not known.

Our observations provide evidence that cardiomyopathy, congestive heart failure and sudden cardiac death play a major role in the clinical course of patients with CDG. In order to describe the heterogeneous clinical presentation of CDG patients affected by cardiomyopathy, we present data on some new cases of childhood CDG associated with cardiomyopathy. Considering the incidence of sudden death in children with this defect, the description of individual courses and management strategies will be of importance for further investigation. We show that early-onset cardiomyopathy is prevalent in infants and children with CDG, that different types of cardiomyopathy are involved, and that cardiomyopathy may be progressive in nature, with significant contribution to morbidity and mortality, including sudden cardiac death. Cardiomyopathy may be the first and for a long time the only manifestation of these disorders. It is therefore obligatory that cardiologists as well as metabolic specialists consider CDG in children presenting with cardiomyopathy.

9.2 Case reports

The screening test for CDG is the isoelectric focusing of serum transferrin. Transferrin has two N-linked carbohydrate chains with 4 terminal negatively charged sialic acids. Structurally abnormal or missing oligosaccharides result in a loss of negatively charged residues, thereby altering the mobility of the molecule in the electric field. All patients had an abnormal IEF test as shown in Figure 1.

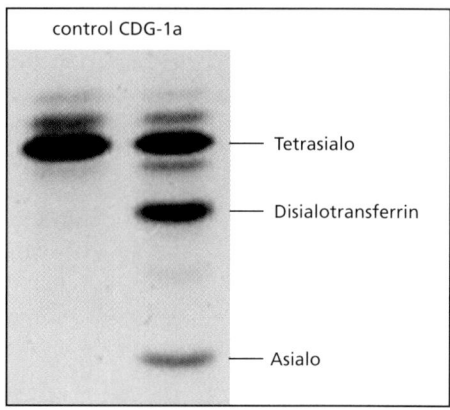

Fig. 1: **Isoelectric focusing of serum transferrin.** Transferrin is a glycoprotein with two N-linked oligosaccharide chains, each carrying two terminal negatively charged sialic acid residues. The majority of transferrin in a control is composed of tetrasialotransferrin. In CDG patients, structural changes in the oligosaccharide chain or complete loss of one or two chains results in a loss of sialic acids thus altering the mobility of transferrin and other glyvoproteins in the electric field. Shown here is a typical patient with CDG-Ia. Transferrin IEF is the typical standard test in order to screen for the presence of CDG and can be performed on a small serum sample. For more information, please see our web site (http://cdg.uni-muenster.de/).

Case 1: CDG-Ia - hypertrophic cardiomyopathy

HJL was the first child of healthy parents. Prenatally, hydrops fetalis with pericardial effusion, ascites and bilateral pleural effusions was diagnosed. Infectious disease screening was negative. The boy was delivered by caesarian section (maternal HELLP syndrome) at 32 weeks of gestation. Due to respiratory insufficiency, intubation and mechanical ventilation were started postnatally and more than 150 ml of ascites fluid were drained. Postpartal physical examination showed generalized oedema and liver enlargement. Recurrent hypoglycaemia occurred. The echocardiogram on the first day of life revealed a small pericardial effusion of 3 mm in systole, but no structural abnormalities, particularly no signs of hypertrophic cardiomyopathy. On day 13 of life, moderate asymmetric septal hypertrophy was noticed, the interventricular septum measuring 6 mm (reference value 3.5-5.5 mm) without left ventricular outflow tract gradient. Two weeks later the pericardial effusion had increased to 9 mm in systole, and now the typical features of hypertrophic cardiomyopathy with pronounced asymmetric septum hypertrophy of 9.2 mm (reference value 4-5.5 mm) with outflow tract obstruction were evident in the echocardiogram (Fig. 2). The Doppler pressure gradient across the left ventricular outflow tract measured 48 mm Hg, which further increased to a constant value of 70 mm Hg 3 days later. The electrocardiogram demonstrated left ventricular hypertrophy with ST segment depression and T wave inversion in the left precordial leads but no arrhythmia. On day 45 of life, a pericardial drainage catheter was inserted surgically as the effusion continued to increase. After the procedure volume requirements were high to maintain adequate cardiac output and epinephrine (0.08-0.2 µg/kg/min) had to be administered. 10 hours later the pericardial line was removed, but the baby remained in low cardiac output state. Some hours later the infant's condition rapidly deteriorated with death occurring from cardiac failure. Post-mortem examination confirmed the clinical cardiac diagnosis. Gross pathology revealed the characteristic morphologic abnormalities of hypertrophic cardiomyopathy including severe asymmetric hypertrophy of the interventricular septum causing outflow tract obstruction and concentric hypertrophy. Histopathologic findings included cardiac muscle cell disorganisation (disarray), fibrosis, and myocyte hypertrophy. Surprisingly, the autopsy revealed significant epicardial bleeding caused by draining of the pericardial effusion, which may have been the ultimate cause for the rapid cardiac deterioration and death of the infant. Coagulation abnormalites are a common feature of CDG. Biochemically, there was no detectable enzyme activity of phosphomannomutase 2 (PMM 2) in the patient's fibroblasts, thus confirming the diagnosis of the most frequent form of CDG, CDG-Ia. DNA analysis showed compound heterozygosity for the paternal 357 C>A mutation in exon 5 and the maternal mutation IVS 3 255+2 T>C.

Case 2 : CDG-Ia - hypertrophic cardiomyopathy [10]

The girl (ByA) was born at term with normal weight, length and head circumference. Muscular hypotonia and inverted nipples were noted. Internal strabismus, a common clinical sign of CDG-Ia, was not present during the first 2 months

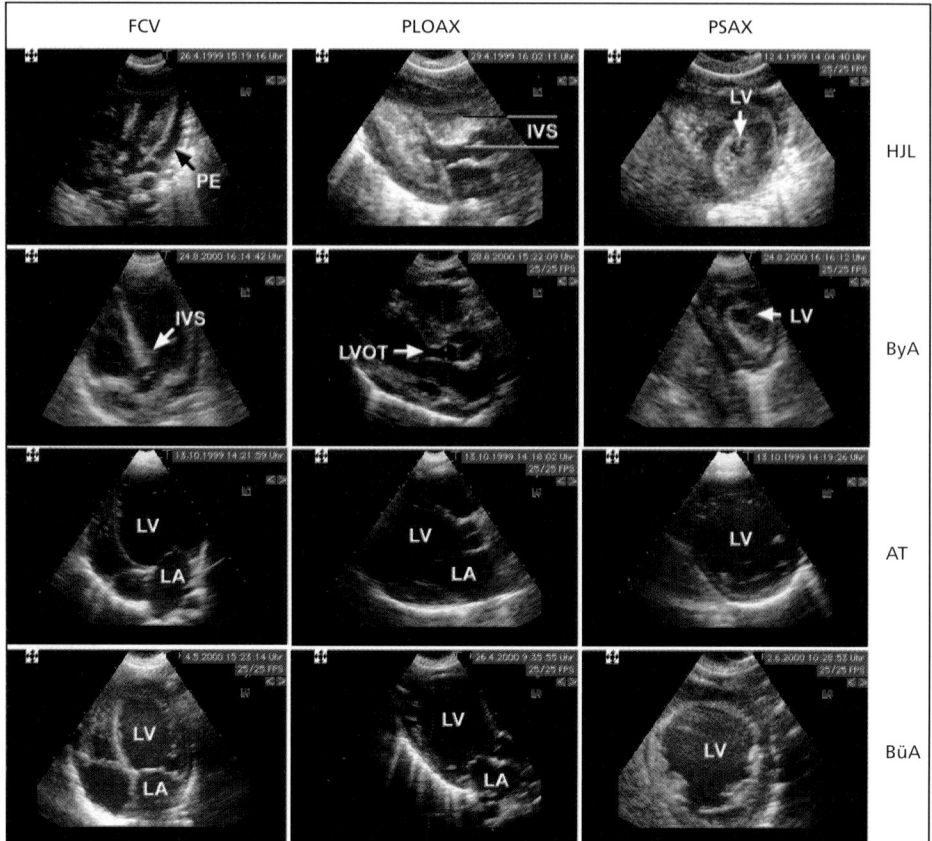

Fig. 2: Representative characteristic echocardiographic features of hypertrophic (patients HJL, ByA, row 1 and 2) and dilated cardiomyopathy (patients AT, BüA, row 3 and 4). (Patient initials on the right of image).
Left panel: apical four chamber view (FCV):
Patient HJL, row 1: pericardial effusion (PE) (arrow) and increased posterior wall thickness
Patient ByA, row 2: asymmetric septal hypertrophy (see IVS - interventricular septal thickness)
Patients AT and BüA, rows 3 and 4: left ventricular (LV) and left atrial (LA) dilatation, note the bulging interventricular septum
Middle panel: parasternal long-axis view (PLAX)
Patient HJL, row 1: left ventricular hypertrophy with marked asymmetric septal hypertrophy (see IVS - interventricular septal thickness), left ventricular outflow tract obstruction, increased posterior wall thickness and pericaridial effusion
Patient ByA, row 2: asymmetric septal hypertrophy and the narrowing of the left ventricular outflow tract (LVOT)
Patients AT and BüA, rows 3 and 4: left ventricular and atrial dilatation
Right panel: paraternal short-axis view (PSAX)
Patients HJL and ByA, rows 1 and 2: concentric left ventricular hypertrophy, very small left ventricular cavity (see LV)
Patients AT and BüA, rows 3 and 4: left ventricular (LV) dilatation

of life but developed later. In the first 8 weeks of life, 3 sepsis-like events occurred without isolation of bacteria in repeated blood cultures. During these episodes, serum transaminase concentrations increased to several hundred units per litre and AT III activity decreased below the detection level. Hypoalbuminaemia and low cholinesterase levels were constantly present. Cerebellar hypoplasia and generalized brain atrophy were detected on MRI. Mild to moderate pericardial effusion was seen on routine echocardiography performed at 2 months of age and persisted thereafter. There was no evidence of structural or functional heart disease at this time. An extended abdomen with loose and frequent stools and inappropriate weight gain led to parenteral nutrition which was continued until 6 months of age. Asymmetrical septal hypertrophy of the left ventricle without outflow tract obstruction was first noted at 6 months of life and was unchanged on follow-up 2 months later. Body weight and length were below the 3rd percentile and weight gain continued to be insufficient even though the child was fed via gastric tube and caloric intake was increased to more than 150 kcal/kg body weight daily. Severe ascites was the major clinical problem for the next months and diuretic therapy was initiated including furosemide, spironolactone and hydrochlorothiazide. CDG diagnostics revealed a type I hypoglycosylation pattern of transferrin. In the patient's leukocytes, phosphomannomutase (PMM) activity was undetectable confirming the diagnosis of CDG-Ia. Two mutations were found in the PMM2 gene: the maternally inherited deletion of 24delC in exon 1 leading to a frame shift and the paternally inherited missense mutation 691 G>A in exon 8 leading to amino acid exchange V231M. At 10 months of age, hypertonic dehydration with a weight loss of 7 % of body weight developed, which was triggered by fever, diarrhoea and ongoing diuretic therapy. Upon arrival in hospital, the child had severe metabolic acidosis (pH 6.6, pCO_2 39 mm Hg, BE –33.1 mmol/l). Serum sodium was 157 mmol/l, creatinine had increased to 1.1 mg/dl and leukocytosis with 61 400 cells/µl was noted. C-reactive protein was below the detection level and IL-8 levels were in the normal range. Body temperature was 39.2° C. Echocardiography now revealed the characteristic feature of hypertrophic cardiomyopathy with marked asymmetrical thickening of the interventricular septum and the left ventricular posterior wall as well as severe left-ventricular outflow tract obstruction (maximum Doppler pressure gradient of ≈ 100 mm Hg) (Fig. 2). The interventricular septum (IVS) measured 11.8 mm (normal value 4-5.5 mm) due to asymmetric thickening, the posterior left ventricular wall (LVPW) 6.2 mm (normal value 4-5 mm). The electrocardiogram showed significant ST elevations as high as 2.5 mV in the left precordial leads compatible with the acute phase of severe myocardial ischaemia. ECG changes within the next few days were similar to those seen after myocardial infarction but normalized much faster. Troponin levels peaked at 9.5 ng/ml (normal < 0.1 ng/ml) but creatinine kinase concentrations remained within the normal range. Mechanical ventilation was started, metabolic acidosis was treated, and the patient was rehydrated. For relief of left ventricular outflow tract obstruction, propranolol at a dose of 0.1 mg/kg four times a day was administered intravenously, which proved to be highly effective. The left ventricular outflow tract gradient decreased from ≈ 100 mm Hg to ≈ 20 mm Hg within 24 hours followed by a marked improvement in the patient's cardiovascular status with complete clinical recovery.

On follow-up at 26 months of age, physical examination of the cardiovascular system was unremarkable. On echocardiography, mild residual asymmetric mid-septal thickening (6 mm) without left ventricular outflow tract obstruction was still present. Whereas systolic left ventricular function was normal (shortening fraction 43 %), diastolic function remained moderately depressed. Beta-blocker therapy was discontinued at that time and on follow-up visit at 31 months of age no recurrent LVOTO was detectable.

Case 3: CDG-x - dilated cardiomyopathy

The girl (BüA) was the first child of consanguineous parents and born after an uneventful pregnancy. Feeding difficulties and poor weight gain were prominent directly after birth. At 7 weeks of age, signs of congestive heart failure became apparent when the infant presented with tachypnoea, feeding difficulties, failure to thrive and significant hepatomegaly. The diagnosis of dilated cardiomyopathy was made echocardiographically. The left ventricle and left atrium were markedly enlarged with the left ventricular diastolic dimension (LVDD) measuring 32 mm (reference value 22-27 mm), and a marked reduction in the shortening fraction (10-17 %) was noted on serial measurements (Fig. 2).

Diagnostic workup at this time excluded any known cause for the cardiomyopathy. Supportive medical treatment including digoxin, diuretics, vasodilators and anticoagulants was instituted, but despite therapy no improvement in cardiac function could be achieved over the following months. Alopecia developed and was complete at 5 months of age. In addition, severe ichthyosis involving the whole body surface was present with histological examination of the skin revealing unspecific hyperkeratosis not attributable to any known disorder. Muscular hypotonia became a prominent clinical feature. Serum cholesterol was low with 54 mg/dl and was mainly due to a very low LDL of 9 mg/dl, whereas HDL was within the normal range (37 mg/dl). Liver transaminases were slightly elevated. Antithrombin III was severely reduced (20 %), coagulation factor XI was reduced to 40% and protein C to 5%. Nerve conduction velocities were unaffected. Isoelectric focusing of serum transferrin revealed a type I pattern (Fig. 1). Analysis of lipid-linked oligosaccharides derived from fibroblasts after metabolic labelling with 3H-mannose showed no structural abnormalities of the dolichol-linked oligosaccharides. At 7 months of age the infant died suddenly while being carried in a stroller. Consent for a post-mortem examination was not obtained.

Cases 4 and 5: CDG-x - dilated cardiomyopathy

The following two patients are siblings. The girl (AT), now 10 years old, was born at term after an uneventful pregnancy with a normal birth weight. Facial dysmorphic features, a cleft palate, hypertelorism, and additional tissue at the left ear were noted postnatally. Body growth was along the 3rd percentile. Psychomotor and mental development were undisturbed and motor nerve conduction velocities were within the normal range. At 16 months of age, in the post-operative period after cleft-palate repair, ventricular arrhythmias and cardiac arrest occurred, necessitating cardio-pulmonary resuscitation. Emergency echocardiography now identified the typical features of dilated cardiomyopathy showing very severe left ventricular dilatation, marked LV dysfunction, and moderate to severe mitral regurgitation. Despite institution of proper treatment, cardiomyopathy remained the major clinical problem over the following years with recurrent hospitalisations for congestive heart failure. Repeated echocardiograms documented massive left ventricular (LVDD 60 mm; normal 31-39 mm), and left atrial (LA/AO ratio 1.9) dilatation and persistent low cardiac function (shortening fraction 7-11 %) (Fig. 2). Endomyocardial biopsy at 6 years of age revealed the full morphological features of dilated cardiomyopathy including myocyte hypertrophy and extensive areas of interstitial fibrosis without significant lymphocytic infiltrate. Non-specific ultrastructural changes were seen on electron microscopy. As there was neither evidence of acute or chronic myocarditis by polymerase-chain-reaction for cardiotropic viruses nor evidence of known metabolic disease, the aetiology of the cardiomyopathy remained unclear. Despite appropriate medical therapy her cardiac status remained severely compromised and she is now listed for heart transplantation. Elevated transaminase levels were noted on several occasions and reduction of serum antithrombin III, thyroxin binding globulin and haptoglobin levels prompted testing for a potential CDG despite absence of other typical features of the syndrome. The presence of a CDG was confirmed, but biochemical analysis excluded all known types.

Her brother (AM), now 16 years old, had a cleft palate as well, but did not exhibit additional dysmorphic features or other physical abnormalities. Initially his clinical course was notable for recurrent episodes of mild hypoglycaemia which resolved. At 7 years of age his echocardiogram, performed to rule out cardiomyopathy because of his sister's cardiac affection, confirmed the presence of borderline dilated cardiomyopathy. His cardiac dysfunction was only mildly to moderately depressed; there was borderline left ventricular dilatation of 41 mm (reference value 31-39 mm) and reduced left ventricular contractility with a shortening fraction of 27%. Clinically he did not experience any symptoms. On regular follow-up his cardiac dimensions and function remain constant at the upper respectively lower limit of normal. Endomyocardial biopsy, performed at age 13, confirmed the diagnosis of early-stage dilated cardiomyopathy, showing beginning endocardial fibroelastosis, small areas of interstitial fibrosis and a slight reduction in the contractile elements. There was no lymphocytic infiltrates and no evidence of acute or chronic myocarditis by polymerase-chain-reaction for cardiotropic viruses. Thus, as in his sister, initially no cause for the cardiomyopathy could be identified. Although his exercise tolerance is decreased to a minor extent with exertional dyspnoea, there has been no deterioration in cardiac function over the past years on repeated echocardiograms. His somatic growth is at the 3rd percentile and he has no mental deficit. In addition to minor elevation of serum transaminases, his serum creatinine kinase levels are slightly elevated. He is treated for hypothyroidism, which is presumably a consequence of reduction in thyroxin binding globulin levels. Serum transferrin, antithrombin III, thyroxin binding globulin and haptoglobin levels were reduced. Biochemical analysis confirmed the presence of a CDG but excluded all known types.

9.3 Discussion

Cardiac involvement in CDG has so far not been adequately recognized in the literature. Taking into account the number of unclear, sudden deaths in young patients with this disorder, this has to change.

Cardiomyopathy is a frequent feature in infants and children with CDG. It contributes significantly to morbidity and mortality, can be progressive in nature, and is associated with a substantial risk of sudden cardiac death. We have documented the reversibility of hypertrophic obstructive cardiomyopathy in one patient and improvement of dilated cardiomypathy in another. In all cases reported so far in the literature, hypertrophic cardiomyopathy was diagnosed and all patients were of the CDG-Ia subtype. However, both dilated and hypertrophic cardiomyopathy were observed in our study population; the hypertrophic form of cardiomyopathy was associated with subtype CDG-Ia in two of our patients, whereas dilated cardiomyopathy was associated with CDG-x in four children.

Our observations indicate that some patients with myocardial involvement may experience life-threatening events caused by extracardiac, disease-related problems (e.g. fever, fluid loss, feeding difficulties) resulting in rapid deterioration of cardiac function. With awareness of the underlying cardiac disease and institution of appropriate and timely medical intervention, resolution, improvement or even prevention of an adverse course can be achieved.

Our observations emphasize the need to consider CDG in the evaluation of genetic conditions associated with cardiomyopathy. We recommend that patients with this disease should undergo routine cardiac examination. Given the high variability of the clinical phenotype in patients with CDG with sometimes subtle clinical features, it is important to realize that cardiomyopathy occasionally may dominate the clinical picture, the first symptom of the disease being cardiac disease. CDG should be included in the differential diagnosis of every child with cardiomyopathy.

Whereas the first CDG was discovered 20 years ago [5, 6], it was only recently that the association of this disease with cardiomyopathy has been recognized [2]. Five cases have been reported in the literature so far (see Table 1, cases 7-11) but only in one case were sufficient data provided. All patients known to date are summarized in Table 1. It is evident that presently reliable data on the frequency or the clinical course of cardiomyopathy in CDG are not available. This may lead to an underestimation of the prevalence of myocardial involvement and may also impede comparisons among future studies. It cannot be excluded that some reported «unclear« deaths or those that occurred under conditions seemingly unrelated to cardiac disease may have had a cardiac cause due to adverse haemodynamic alterations imposed upon a hitherto unrecognized preexisting cardiomyopathy.

We suggest that all patients with CDG should be screened for cardiomyopathy, even in the absence of clinical signs and symptoms which may be initially subtile. Echocardiography is the principle

Cardiomyopathy in congenital disorders of glycosylation (CDG) 95

Case	Patient/ Ref. #	CDG Type	Biochemical Basis	Genotype	Cardiac Phenotype	Sex	Age at diagnosis	Outcome cardiac	Cause of Death
1	H.J.L.	CDG-Ia	PMM 2	Compound heterozygosity for the paternal 357 C>A mutation in exon 5 and the maternal mutation IVS 3 255+2T>C.	Hypertrophic cardiomyopathy (HCM), Pericardial effusion Echo: asymmetric septal hypertrophy IVS = 9.2 mm; LVOTO: Doppler gradient max. 70 mm Hg. Pathology: septal thickening, disarray of myocardial fibres	M	2 weeks	Death on day of life 47	Cardiac failure; epicardial bleeding (iatrogenic)
2	By.A.	CDG-Ia	PMM2	Mat. inherited deletion of 24delC; pat. inherited missense mutation 691 G>A in exon 8.	Hypertrophic cardiomyopathy (HCM), Pericardial effusion Echo: asymmetric septal hypertrophy, IVS = 11.5 mm; LVOTO: Doppler gradient max. 100 mm Hg.	F	6 months	Alive resolution of LVOTO (at 26 months: IVS = 5.6 mm); some residual diastolic dysfunction	
3	Bü.A.	CDG-x	unknown	-	Dilated cardiomyopathy (DCM) Echo: LV dilatation: LVDD = 32 mm LV dysfunction: SF = 8-15 %	F	3 months	Sudden death at 7 months of age	Sudden death
4	T.A.	CDG-x	unknown	-	Dilated cardiomyopathy (DCM) Echo: LV dilatation: LVDD = 60 mm LV dysfunction: SF = 7 - 11 % Pathology: myocyte hypertrophy, interstitial fibrosis	F	8 months	Alive severe cardiac dysfunction listed for cardiac transplantation	
5	M.A.	CDG-x	unknown	-	Dilated cardiomyopathy (DCM) Echo: LV dilatation: LVDD = 41 mm LV dysfunction: SF = 27% Pathology: myocyte hypertrophy, interstitial fibrosis	M	7 years	Alive minor cardiac dysfunction	
6	N.H.	CDG-x	unknown	-	Dilated cardiomyopathy (DCM) Echo: LV dilatation: LVDD = 44 mm LV dysfunction: SF = 22 %	F	5 months	Alive improvement, but some residual cardiac dysfunction	
7	5	CDG-Ia	PMM2	?	Hypertrophic cardiomyopathy (HCM), Pericardial effusion Echo: «left and right ventricular hypertrophy of 0.5 cm«; stiff ventricle« Left ventricular mid-cavity obstruction: Doppler gradient: 80 mm Hg	M	3 weeks	Death at 11 weeks of age	Rapidly, progressive obstructive hyper trophic cardiom yopathy, extremely poor weight gain, apnea
8	10	?	?	?	Hypertrophic cardiomyopathy (HCM)? Echo: «some thickening of the walls of both ventricles« Pathology: «thickened ventricular walls«	M	3 months	Death at 15 weeks of age	«Suddenly became restless with an irregular breathing pattern»
9	12	CDG-I?	?	?	Hypertrophic cardiomyopathy (HCM), Pericardial effusion Echo: - no data presented - Pathology: « hypertrophic non-obstructive cardiomyopathy«	M	prenatal	Death on day of life 42	Septicemia and meningitis
10	16	CDG-Ia	PMM 2	R141H	Hypertrophic cardiomyopathy (HCM) Echo: - no data presented -	M	2 years	?	?
11	13	CDG-Ia	PMM 2	R141H/F119L	«Poor ventricular function» Pleural effusion Echo: - no data presented	M	4 months	Death at 6 months of age	?

Tab. 1: Synopsis of Clinical, Biochemical, Genetic and Cardiac Data (Literature Re-view/Our Cases) IVS = interventricular septum; LVDD = left ventricular diastolic dimension; SF = shortening fraction; LVOTO = left ventricular outflow tract obstruction; PMM = phosphomannomutase

diagnostic modality with the highest sensitivity and specificity. Echocardiographic criteria for abnormal left ventricular systolic performance and structure in infants and children have been established [3, 4]. For the diagnosis of hypertrophic cardiomyopathy, the measurement of septal thickness based on weight is used in children for diagnosis [15]. The use of standardized, objective criteria improves not only the accuracy of the diagnosis but also the understanding of the clinical course of cardiomyopathy [14] in these newly recognized disorders. As aetiology-specific treatment might in future be implemented successfully in some patients with CDG [7, 9, 12], it is highly important to make a timely diagnosis [8].

Conversely, any patient with cardiomyopathy who presents with additional signs and symptoms of multiple organ dysfunction, such as neurological (hypotonia, strabismus, ataxia), cutaneous (inverted nipples, abnormal distribution of adipose tissue) abnormalities, or multivisceral involvement (digestive, hepatic, renal) should be screened for CDG. Testing for CDG can be performed fast, reliably und inexpensively in an experienced laboratory.

It can be speculated that the general hypoglycosylation of glycoproteins in this syndrome may induce alterations of the dystrophin-associated glycoproteins in the sarcolemmal plasma membrane which ultimately leads to cardiomyopathy. It remains unclear whether a common pathogenic mechanism is involved leading to the two manifestations of hypertrophic and dilative cardiomyopathy. The «final common pathway« hypothesis suggests different aetiologies for these two forms of cardiomyopathy [15]. Whereas hypertrophic cardiomyopathy is a disease of the sarcomere, dilated cardiomyopathy is associated with abnormalities in cytoskeletal proteins [1]. Tissue-specific protein expression and/or alterations in the dystrophin-associated glycoproteins that play a major role in maintaining integrity and mechanical properties of the plasma cell membrane and other critical functions, such as signal transduction and calcium homeostasis, seem to be involved in the cascade of events that ultimately lead to dilated cardiomyopathy. However, two lines of experimental evidence at the molecular level seem to support the observation of the occurrence of two different functional myopathic manifestations in the same group of disorders. In an animal model of disrupted dystrophin-associated glycoprotein complex, a defect in a gene for δ-sarcoglycan was found to cause both hypertrophic and dilated cardiomyopathy [13]. In humans, a mutation in the sarcomeric α-cardiac actin gene was found to be responsible for familial hypertrophic cardiomyopathy and for idiopathic dilated cardiomyopathy [11]. One way to test this hypothesis would be to study the glycosylation pattern in myocardium of affected patients. To our knowledge this has not yet been undertaken.

9.4 Conclusion

Cardiomyopathy – either dilated or hypertrophic – seems to be an inherent component of CDG over a range of subtypes and places infants and children with this disorder at a substantial risk for sudden cardiac death. Therefore, cardiomyopathy should be considered in any patient with this disorder and likewise CDG should be included in the differential diagnosis in paediatric patients with «idiopathic» cardiomyopathy. Although CDG are heterogeneous diseases characterized by a wide clinical and biochemical heterogeneity among different types, cardiomyopathies do occur in different types of the syndrome, but apparently related to the biochemical defect as evidenced by our preliminary data. Further pooling of reported cases associated with cardiomyopathy and phenotype–genotype correlations of the different CDG subtypes are critical to fully appreciate the spectrum of cardiac involvement, and to better define the temporal evolution and outcome of cardiomyopathy in this syndrome. The cardiac pathophysiology in CDG may be explained in the future once the potential involvement of the dystrophin glycoprotein complex in this disease is better understood.

References

[1] Badorff, C., G. H. Lee, B. J. Lamphear, M. E. Martone, K. P. Campbell, R. E. Rhoads, K. U. Knowlton: Enteroviral protease 2A cleaves dystrophin: evidence of cytoskeletal disruption in an acquired cardiomyopathy. Nat. Med. **5**, 320-326 (1999).

[2] Clayton, P. T., B. G. Winchester, G. Keir: Hypertrophic obstructive cardiomyopathy in a neonate with the carbohydrate-deficient glycoprotein syndrome. J. Inher. Metab. Dis. **15**, 857-861 (1992).

[3] Colan, S. D., I. A. Parness, P. J. Spevak, S. P. Sanders: Developmental modulation of myocardial mechanics: age- and growth-related alterations in afterload and contractility. J. Am. Coll. Cardiol. **19**, 619-629 (1992).

[4] Henry, W. L., J. M. Gardin, J. H. Ware: Echocardiographic measurements in normal subjects from infancy to old age. Circulation **62**, 1054-1061 (1980).

[5] Jaeken, J., H. G. van Eijk, C. van der Heul, L. Corbeel, R. Eeckels, E. Eggermont: Sialic acid-deficient serum and cerebrospinal fluid transferrin in a newly recognized genetic syndrome. Clin. Chim. Acta **144**, 245-247 (1984).

[6] Jaeken, J., M. Vanderschueren-Lodeweyckx, P. Casaer, L. Snoeck, L. Corbeel, E. Eggermont, R. Eeckels: Familial psychomotor retardation with markedly fluctuating serum prolactin, FSH and GH levels, partial TBG deficiency, increased serum arylsulphatase A and increased CSF protein: a new syndrome? Pediatr. Res. **14**, 179 (1980).

[7] Lühn, K., T. Marquardt, E. Harms, D. Vestweber: Discontinuation of fucose therapy in LAD II causes rapid loss of selectin ligands and rise of leukocyte counts. Blood **97**, 330-332 (2001).

[8] Marquardt, T., H. H. Freeze: Congenital disorders of glycosylation. Glycosylation defects in man and biological models for their study. Biol. Chem. **382**, 161-177 (2001).

[9] Marquardt, T., K. Lühn, G. Srikrishna, H. H. Freeze, E. Harms, D. Vestweber: Correction of leukocyte adhesion deficiency type II with oral fucose. Blood **94**, 3976-3985 (1999).

[10] Marquardt, T., G. Hulskamp, J. Gehrmann, V. Debus, E. Harms, H. G. Kehl: Severe transient myocardial ischaemia caused by hypertrophic cardiomyopathy in a patient with congenital disorder of glycosylation type Ia. Eur. J. Pediatr. **161**, 524-527 (2002).

[11] Mogensen, J., I. C. Klausen, A. K. Pedersen, H. Egeblad, P. Bross, T. A. Kruse, N. Gregersen, P. S. Hansen, U. Baandrup, A. D. Borglum: Alpha-cardiac actin is a novel disease gene in familial hypertrophic cardiomyopathy. J. Clin. Invest. **103**, R39-43 (1999).

[12] Niehues, R., M. Hasilik, G. Alton, C. Körner, M. Schiebe-Sukumar, H. G. Koch, K.-P. Zimmer, R. Wu, E. Harms, K. Reiter, K. von Figura, H. H. Freeze, H. K. Harms, T. Marquardt: Carbohydrate-deficient glycoprotein syndrome type Ib. Phosphomannose isomerase deficiency and mannose therapy. J. Clin. Invest. **101**, 1414-1420 (1998).

[13] Sakamoto, A., K. Ono, M. Abe, G. Jasmin, T. Eki, Y. Murakami, T. Masaki, T. Toyo-oka, F. Hanaoka: Both hypertrophic and dilated cardiomyopathies are caused by mutation of the same gene, delta-sarcoglycan, in hamster: an animal model of disrupted dystrophin-associated glycoprotein complex. Proc. Natl. Acad. Sci. USA **94**, 13873-13878 (1997).

[14] Schwartz, M. L., G. F. Cox, A. E. Lin, M. S. Korson, A. Perez-Atayde, R. V. Lacro, S. E. Lipshultz: Clinical approach to genetic cardiomyopathy in children. Circulation **94**, 2021-2038 (1996).

[15] Towbin, J. A.: Pediatric myocardial disease. Pediatr. Clin. North. Am. **46**, 289-312, (1999).

10 Cardiovascular changes in the mucopolysaccharidoses

C.-F. Wippermann, M. Beck, D. Schranz, R. Huth,
I. Michel-Behnke, B.-K. Jüngst

Mucopolysaccharidoses represent a group of inborn defects characterized by an incomplete intralysosomal degradation of acid mucopolysaccharides (glycosaminoglycans). They are caused by gene mutations which lead to individual, biologically inactive enzymes necessary for the degradation of glycosaminoglycans. Glycosaminoglycans are high molecular weight carbohydrate chains with important functions in different organ systems, especially in connective tissue.

The clinical symptoms (Table 1) are caused as a consequence of the storage of glycosaminoglycans in different organ systems. According to clinical and chemical signs, six main forms and several subtypes of mucopolysaccharidoses can be distinguished. Clinical and biochemical reviews are available [23, 27]. We would like to emphasize at this point that in our opinion the consideration of invasive procedures (e.g. valve replacement) in patients with MPS should take the other, especially mental, symptoms into account. These procedures should only be undertaken after consultation with other physicians taking care of the patient.

As early as 1917, C. H. Hunter wrote in the description of a patient with mucopolysaccharidosis type II and cardiomegaly: "... a distinct diastolic murmur audible in the third and fourth left intercostal spaces close to the sternum...; at the apex, a systolic murmur was conducted towards the axilla" [15]. This demonstrates that cardiovascular alterations may be part of the clinical symptomato-

Tab. 1: The mucopolysaccharidoses.

Syndrome	Type	~ Incidence	Mental retardation	Dysmorphic features	Corneal clouding	Somatic changes
Hurler	IH	1:100000	+++	+++	++	+++
Scheie	IS	1:500000	–	++	++	++
Hunter	II (A, B)	1:150000	variable	+–+++	–	+++
Sanfilippo	III (A-D)	1:24000	+++	+	–	+
Morquio	IV (A, B)	1:330000	–	specific	++	variable
Maroteaux-Lamy	VI (A, B)	rare	–	+++	++	variable
Sly	VII	rare	++	+–++	++	++

logy in mucopolysaccharidoses, and it is the objective of this article to describe their spectrum in these patients. This will be realized mainly in the form of tables that summarize the published results and our own data.

Abbreviations:

MPS: mucopolysaccharidosis
MPS IH: Hurler syndrome
MPS IS: Scheie syndrome
MPS II: Hunter syndrome
MPS III: Sanfilippo syndrome
MPS IV: Morquio syndrome
MPS VI: Lamy-Maroteaux syndrome

For the interpretation of the summarized results taken from different authors and studies, the following factors have to be taken into consideration:
- in most cases there are only small numbers of patients or individual case reports.
- very often only selected patient groups (e.g. cardiological patients) were investigated without comparison with a control group.
- the extent of the disease problem is age dependent, that means older children are more frequently described than infants.
- diagnosis of patients in the early studies was normally not based on enzymatic analysis, therefore diagnostic errors cannot be excluded.
- different diagnostic cardiological techniques and procedures (i.e. auscultation, echocardiography, catheter) with different levels of diagnostic accuracy were employed.

Thus, all the above aspects limit the formulation of generalized statements.

10.1 Changes of the coronary arteries

10.1.1 Pathoanatomical results

Table 2 presents the publications dealing with cardiovascular problems mainly in patients with MPS I and MPS II [3, 5, 20, 28, 34]. Relevant stenoses of the extramural coronary arteries were found in 34 out of 82 patients. These stenoses are essentially caused by an intima thickening on the basis of numerable typical storage cells, called "Hurler cells". To a lesser extent these storage cells are also found in the media and in the intramural coronary arteries. With

Tab. 2: Anatomical/pathological changes in coronary arteries.

Author	Reference	MPS patients	Relevant stenoses	Ischaemia (clinical)	Ischaemia (ECG)	Myocardial infarction
Krovetz	[20]	Literature	24/68	?	?	0/68
Young	[34]	MPS II	1/4	?	?	0/4
Brosius	[5]	MPS I	5/6	0/6	0/6	0/6
Stephan	[28]	Cardiomyopathies	2/2	0/2	1/2	0/2
Belani	[3]	Peri-OP	2/2	1/2	1/2	1/2
	Total		34/82 (41.5%)	1/10	2/10	1/82 (1.4%)

electron microscopy "laminar bodies" consisting of concentric or parallel lamellae 35 Å broad and with individual distance of 30 Å, were additionally described [24, 25]. Despite significant stenoses, infarction and ischaemia were only reported individually in one patient. A detailed description of the histological changes is given by Brosius et al. [5] and Renteria et al. [25].

The severity of the stenoses and their pathoanatomical basis was quantitatively investigated by Brosius et al. [5]. The 6 patients with MPS I never presented clinical or electrocardiographic signs of cardiac dysfunction or myocardial ischaemia. However, the extent of coronary stenoses described by Brosius et al. was equal to that in patients dying from coronary heart disease [5].

10.1.2 Clinical results

25 patients with MPS investigated by coronary angiography are described in the literature. Six of them showed a significant stenosis of at least one coronary artery (Table 3) [3, 6, 20, 29]. It is suggested that the coronary stenoses are underestimated by angiography, because one patient dying during coronary angiography presented anatomically with more severe stenoses than estimated by angiography [4]. In 2 of 138 patients signs of coronary ischaemia were clinically apparent. However, the mentally retarded patients may have been unable to articulate the symptoms of myocardial ischaemia. In only one patient did the ECG present signs of myocardial ischaemia and infarction [3].

10.1.3 Discussion

Presumably the MPS are diseases with the most extensive forms of stenosis of the coronary arteries observed during childhood [5]. More remarkable is the extreme discrepancy between pathoanatomically demonstrable extreme stenoses and lacking clinical symptoms. The cause of this controversy has not been elucidated. One explanation can be given on the basis of the following spec-

Tab. 3: Coronary artery alterations in MPS patients (VR = Valve replacement).

Author	Reference	MPS patients	Angiographically relevant stenoses	Ischaemia (clinical)	Ischaemia (ECG)	Myocardial infarction	Sudden death
Craig	[7]	MPS I	?	1/2	?	?	?
Schiecken	[26]	Cardiol.	?	0/9	0/9	0/9	?
Young	[34]	MPS II	?	0/31	?	0/31	3/13
Nelson	[22]	Unselected	?	0/10	0/10	0/10	?
John	[16]	MPS IV	?	0/22	0/22	0/22	?
Belani	[3]	Peri-OP	5/12	1/30	1/30	1/30	1/30
Krovetz	[20]	Cardiol.	1/11	0/32	0/32	0/32	10/87
Butman	[6]	VR, MPS IS	0/1	0/1	0/1	0/1	?
Tan	[24]	VR, MPS VI	1/1	0/1	0/1	0/1	?
Total			6/25 (24%)	2/138 (1.4%)	1/105 (1%)	1/136 (0.7%)	14/130 (11%)

ulation. In coronary heart disease endothelial dysfunction in the arteriosclerotic vessel wall plays an important role [9] supporting the formation of local thrombosis. Especially when fissures and ruptures of an arteriosclerotic plaque occur, their thrombogenetic contents enter the vessel lumen and induce thrombus formation. In the MPS of major concern (MPS I, II, VI), the vascular stenoses are caused by the accumulation of dermatan sulphate and/or heparan sulphate. These, however, have clear antithrombotic activity [12, 21] and therefore can prevent thrombus formation especially in the area of the stenosis as in coronary heart disease.

10.2 Alterations of other arterial vessels

Table 4 summarizes a series of pathoanatomical as well as published clinical results [6, 20, 25, 30, 31]. Pathoanatomically the alterations correspond to those seen in the coronary arteries [25]. In 30 of 100 patients alterations of the aorta were observed; in 7% of the patients the alterations were so excessive that symptoms of aortic coarctation stenosis with decreased femoral pulses were observed [2, 30, 31]. The stenoses, however, were most frequently observed in the abdominal aorta. In addition other stenoses, e.g. in the renal artery, were observed. Angiographies are published by Taylor et al. [30].

10.3 Arterial hypertension

In stenoses of the aorta or the renal artery the occurrence of arterial hypertension is not unusual. As shown in Table 5, 31 of 101 patients presented with arterial hypertension of unclear origin apart from the described vascular changes [18, 20, 22, 25-30, 31]. In some cases the observed hypertension was excessive with a systolic blood pressure between 200 and 250 mm Hg [20].

Tab. 4: Changes in other arterial vessels (Coa = Symptoms similar to aortic coarctation).

Author	Reference	MPS patients	Aorta	Other vessels
Krovetz	[20]	Literature patholog.-anatom.	18/68	5/68
Renteria	[24]	MPS IH, patholog.-anatom.	5/5	5/5
Beaudet	[2]	MPS VI	1/1 Coa	?
Taylor, D.	[30]	MPS IH/IS	5/24 Coa	5/24
Taylor, J.	[31]	MPS I	1/2 Coa	?
	Total		30/100 7/100 Coa	10/97

Tab. 5: Arterial hypertension.

Author	Reference	MPS patients	Arterial hypertension
Krovetz	[20]	Cardiological	9/32
Krovetz	[18]	Literature pathol.-anatom.	7/22
Renteria	[24]	MPS IH, pathol.-anatom.	2/5
Taylor, J.	[31]	MPS 1, nephrological	2/2
Nelson	[22]	50% MPS IV	3/16
Taylor, D.	[30]	MPS I	8/24
Total			**31/101 (31%)**

10.4 Myocardial changes

10.4.1 Pathoanatomical results

Typical storage cells are found in the myocardial interstitium although not as frequent as in the coronary arteries or valves [24, 25]. In the myocytes vacuoles are occasionally found and small areas of fibrosis have been described. The myocytes are slightly hypertrophied. The proliferation of elastic and collagen like fibrils as well as individual storage cells cause a thickening of the mural endocardium. As demonstrated in Table 6, the endocardium is more frequently affected than the myocardium [8, 10, 20, 25, 28].

10.4.2 Clinical observations

When reviewing the published studies and our own data (Table 7), two groups can be observed. The majority of patients show a normal systolic function, only 18% of the patients listed in Table 7 showed reduced indices of myocardial performance. This was frequently seen in patients with MPS IH, IS, II and VI [11, 16, 19, 20, 22, 26]. In our own investigations a decreased shortening fraction was found in only 2 out of 74 patients. An additional significant num-

Tab. 6: Pathological-anatomical alterations in endocardium and myocardium.

Author	Reference	MPS patients	Myocardium	Endocardium	Cardiac insufficiency (clinical)
Krovetz	[20]	Literature	9/68	14/68	?
Renteria	[24]	MPS IH	5/5	5/5	0/5
Fong	[10]	Cardiomyopathy MPS VI	(1)/1	1/1	1/1
Stephan	[28]	Cardiomyopathy MPS IH	1/2	2/2	2/2
Donaldson	[8]	Cardiomyopathy MPS IH	2/3	2/3	3/3
Total			**17/79 (22%)**	**24/79 (30%)**	

Tab. 7: Myocardial function in MPS patients.

Author	Reference	Method	IH	IS	II	III	IV	VI	Total
Krovetz	[20]	Catheter	6/19	0/0	3/10	0/0	0/1	1/2	10/32
Schiecken	[26]	m-mode	0/0	2/2	3/3	0/1	0/1	1/1	6/8
Johnson	[17]	m-mode	2/4	0/0	0/1	0/0	0/0	0/0	2/5
Gross	[11]	2D-Echo	4/7	0/6	0/7	0/6	0/3	0/0	4/29
John	[16]	2D-Echo	0/0	0/0	0/0	0/0	0/11	0/0	0/11
Nelson	[22]	2D-Echo	1/2	2/3	0/2	2/4	3/11	0/0	8/22
	This study	2D-Echo	0/11	1/6	0/14	0/24	0/11	1/8	2/74
			13/43 (30%)	5/17 (29%)	6/37 (16%)	2/35 (6%)	3/38 (8%)	3/11 (27%)	32/181 (18%)
Cardiomyopathy		Case reports	7/7				3/3		

ber of patients had shortening fractions at the lower limit of normal. However when it is taken into consideration that these patients are suffering from a valvular insufficiency or were very restless, this shortening fraction may be no longer normal. In addition an asymmetrical thickening of the septum was frequently observed. This, however, was not accompanied by obstruction of the outflow tract (Figs. 2 and 5) [11].

In Table 7 a second small group of patients, mostly newborns, is listed who presented with cardiomyopathy as the first symptom of MPS [8, 10, 13, 28]. Either dilated cardiomyopathy or endocardial fibroelastosis were present.

Most of the patients died from cardiac failure, although the cause of the myocardial failure of these patients compared to the majority mentioned above remains unexplained.

10.5 Changes of the conduction system

In one case it could be histologically demonstrated that the structures of the conduction system were altered by stor-

Tab. 8: ECG abnormalities in MPS patients (VES, SVES = ventricular, supraventricular extra systole; SVT = supraventricular tachycardia).

Author	Reference	P-Q lengthening	QTc lengthening	Extra systole	Tachycardia	Sudden death
Krovetz	[20]	?	common	?	?	10/87
Schiecken	[26]	2/9	0/9	1/9 VES	0/9	?
Young	[34]	?	?	?	?	3/13
Gross	[11]	0/18	1/18	1/18SVES	1/18SVT	?
Nelson	[22]	1/22	3/22	0/22	0/22	?
Total		3/49 (6%)	4/49 (8%)	2/49 (4%)	1/49 (2%)	13/100 (13%)

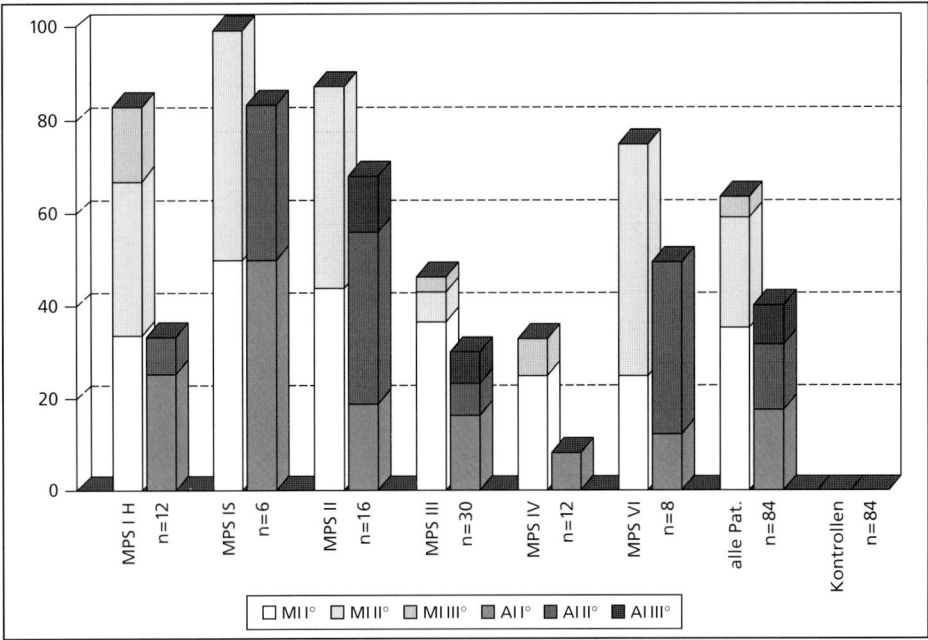

Fig. 1: **Incidence in percent of aortic and mitral insufficiency in 84 patients with mucopolysaccharidosis.** AI, MI I0, II0, III0: mild, middle grade, severe aortic, mitral insufficiency.

age cells. Renteria et al. described a focal accumulation of storage cells between the cells of the AV node [25] and in some patients corresponding prolongations of the PQ time were found (Table 8) [11, 20, 22, 26, 34]. Also prolongations of the QTc time as well as individual ventricular and supraventricular extrasystoles were observed. Because tachycardias were observed only in one case, the clinical consequences of the histologically demonstrable alterations in the conductive system seem to be rare. It cannot be excluded, however, that the relatively frequent sudden deaths in MPS patients are caused by cardiac arrhythmias.

10.6 Alterations of the heart valves

10.6.1 Pathoanatomical results

Macroscopically altered heart valves are seen in patients with MPS (Table 9) [3, 8, 10, 20, 25, 28]. The valves are thickened and at the free borders irregular additional nodules are found (Fig. 3). The mitral, aortic, tricuspid and pulmonary valves are affected in decreasing order. The tendinous cords of the atrioventricular valves are also thickened and shortened. Histologically the thickened areas of the valves correspond to large cells with a clear cytoplasm (storage or "Hurler" cells) and to accumulation of collagenous tissue [24, 25].

Tab. 9: Pathological/anatomical valvular alterations.

Author	Reference	MPS patients	Valvular alterations
Krovetz	[20]	Literature	48/68
Renteria	[25]	MPS I	5/5
Fong	[10]	Cardiompathy MPS VI	0/1
Stephan	[28]	Cardiompathy MPS IH	2/2
Donaldson	[8]	Cardiompathy MPS IH	2/4
Belani	[3]	Peri-OP	1/1
Total			58/81 (72%)

10.6.2 Clinical Findings

The thickened cardiac valves can be demonstrated sonographically [11, 16, 17]. In our own studies of 84 patients excessive thickening of the mitral valve was found in 38% and slight changes in 54% (Figs. 2 and 5). A thickening of the aortic valve was seen in 55%.

Corresponding to the frequent morphological valvular alterations, regurgitations of the valves are frequently found. Presumably MPS is the most frequent cause of valvular regurgitations in childhood. In 52% to 75% of all patients an aortic or mitral regurgitation exists (Table 10 and Figs. 4 and 5) [11, 16, 20, 22].

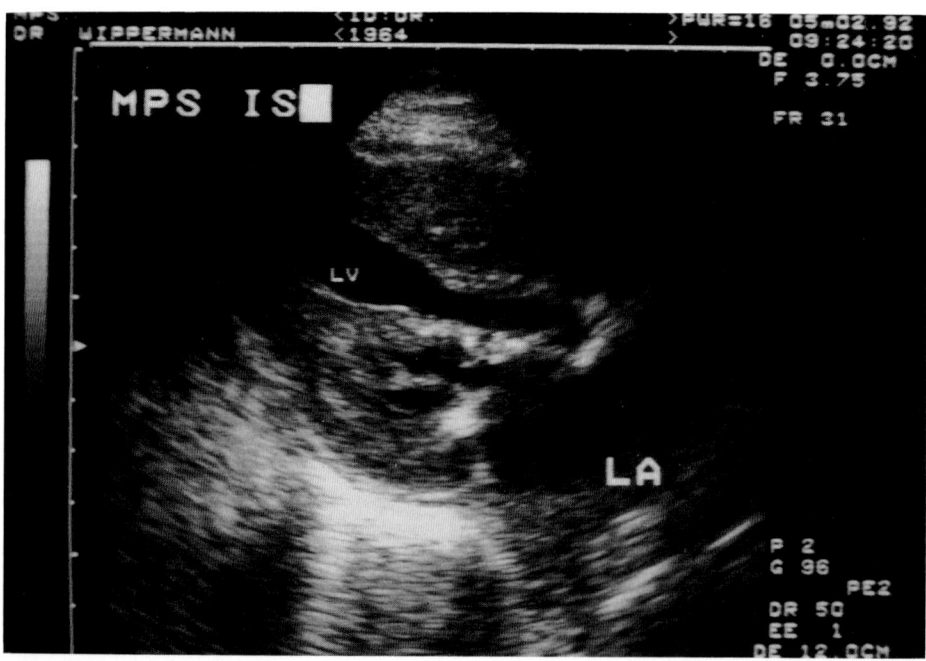

Fig. 2: Echocardiogram in the parasternal long axis in a 30-year-old female with MPS IS and mitral stenosis. Of note is the highly thickened mitral valve with nodular distensions on the valve rim and in the chordae tendinae. The myocardium is also thickened. LA = left atrium; LV = left ventricle.

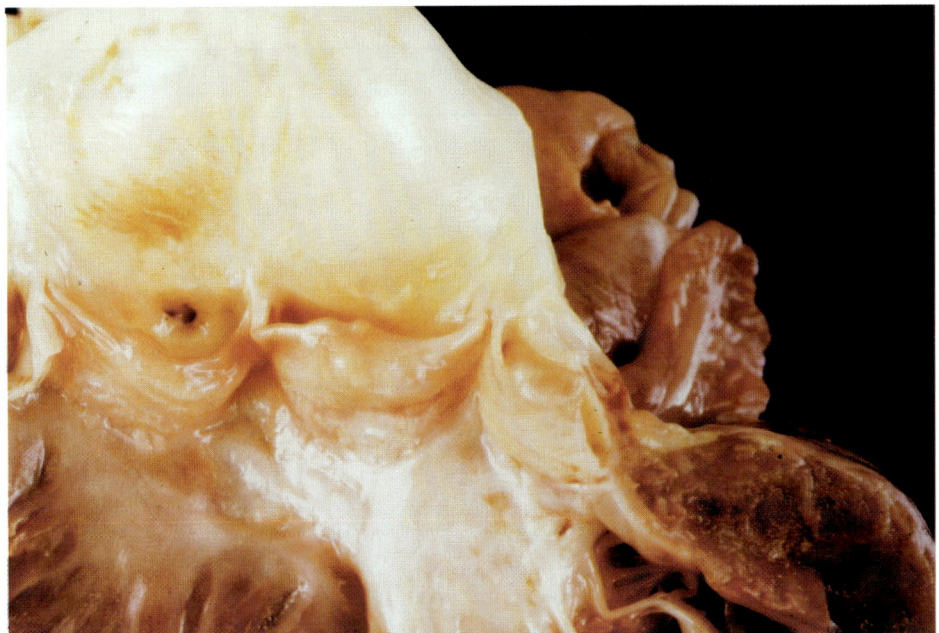

Fig. 3: **Anatomical preparation of the left ventricular outflow tract and ascending aorta.** Nodular distensions are seen on the edge of the cusp (Cortesy of Dr. K. Ulrich, University of Münster).

The reported frequency strongly depends on the method of investigation. This is of clinical relevance because even severe valvular regurgitation may be overlooked when relying on auscultation [11, 16]. Valvular insufficiencies are especially frequent in MPS IS, II and VI and according to our results also in MPS IH. In contrast patients with MPS IV are rarely affected.

As shown in our own investigations most of the insufficiencies are only mild or moderate (Fig. 1). Only 5% of our 84 patients presented with severe mitral insufficiency and 8% with severe aortic regurgitation. Compared to the valvular regurgitations, the stenoses are relatively infrequent. As presented in Table 11, an aortic or mitral stenosis was found in 9% in both groups (Fig. 5) [11, 16, 20, 22, 26]. Among our

Tab. 10: **Aortic and/or mitral regurgitation in MPS patients** (2D-E = 2-dimensional echocardiography).

Author	Method	IH	IS	II	III	IV	VI	Total
Krovetz	Catheter, Auscult.	5/19	0/0	3/10	0/0	0/0	1/2	9/32
Gross	2D-E, Auscult.	4/7	4/6	4/7	0/6	1/3	0/0	13/29
John	2D-E, Doppler	0/0	0/0	0/0	0/0	5/11	0/0	5/11
Nelson	2D-E, Doppler	0/2	0/3	2/2	1/4	0/11	0/0	3/22
This study	Colour Doppler	10/12	6/6	15/16	20/30	4/12	8/8	63/84
	Total	19/40 (48%)	10/12 (83%)	24/35 (69%)	21/40 (53%)	10/37 (27%)	9/10 (90%)	93/178 (52%)

Tab. 11: Aortic and mitral stenosis.

Author	Reference	Method	Aortic stenosis	Mitral stenosis
Krovetz	[20]	Catheter	0/25	0/25
Schiecken	[26]	m-mode	0/9	2/9
Gross	[11]	2D-Echo, Doppler	3/29	0/29
John	[16]	2D-Echo, Doppler	0/11	1/11
Nelson	[22]	2D-Echo, Doppler	1/22	0/22
	This study	2D-Echo, Doppler	3/82	4/82
		Total	7/178(9%)	7/178(9%)
	Case reports	Valve replacement	5/6	5/6

patients were 10 others, besides those 4 with a significant mitral stenosis, who presented with an increased inflow velocity through the mitral valve. Using Doppler echocardiography, no patient was seen with a gradient > 40 mmHg across the aortic valve and > 20 mmHg across the mitral valve. However, when the patients with valvular replacement are analysed, 5 out of 6 had a severe aortic and mitral stenosis [6, 14, 29, 33].

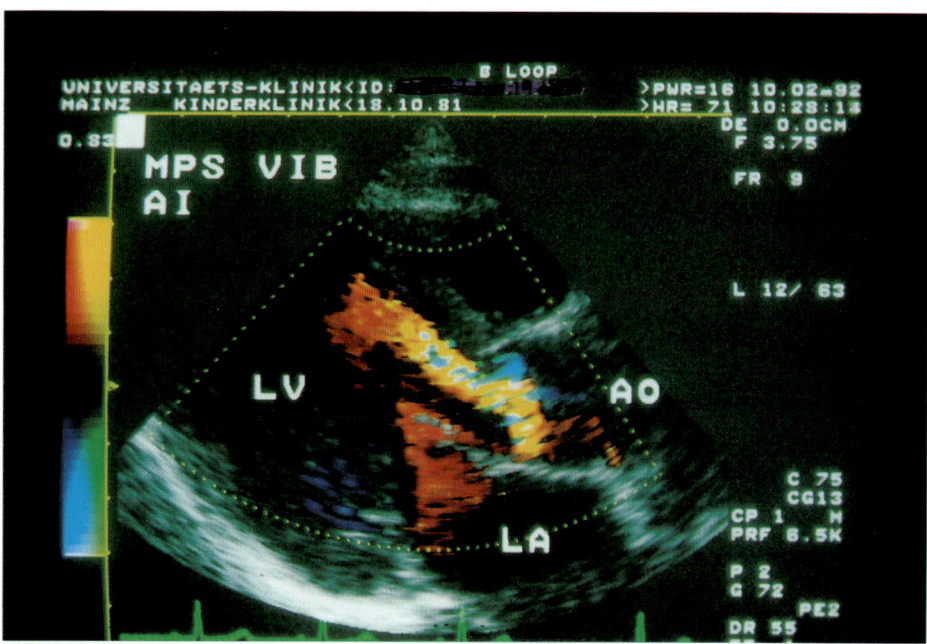

Fig. 4: Colour Doppler echocardiogram in the parasternal long axis in a 12-year-old female with MPS VI and moderate to severe aortic regurgitation. In the diastole the opened mitral valve is seen together with the yellow-turquoise mottled jet of the aortic regurgitation. LA = left atrium; left ventricle; Ao = aorta.

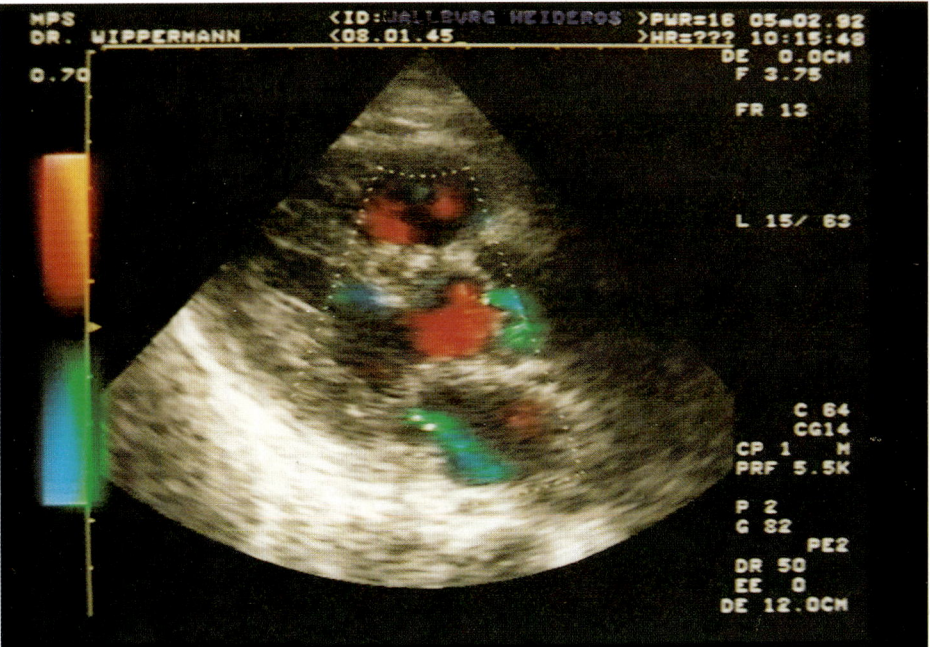

Fig. 5: Colour Doppler echocardiogram in the parasternal long axis in a 38-year-old female with MPS VI and slight mitral regurgitation and mild aortic stenosis. In this systolic picture, the turquoise-coloured jet of the mitral regurgitation in the left atrium is seen under the thickened mitral valve together with turquoise-coloured turbulences in the dilated ascending aorta due to the poorly opened aortic valve. Also present is a slight, non-obstructive asymmetrical septal hypertrophy.

10.7 Therapeutic influence on the cardiovascular alterations

10.7.1 Bone marrow transplantation

There are only few published reports about cardiovascular changes after bone marrow transplantation. Vinallonga reported a patient with MPS I with septum hypertrophy which normalized, decreasing from 11 mm to 5 mm diameter after bone marrow transplantation [32]. On the other hand it was reported that in a child with MPS IH the angiographically proven coronary stenosis could still be demonstrated 13 months after a successful transplant [4].

10.7.2 Cardiac valve replacement

Replacement of the cardiac valves has been successfully performed [6, 14, 29, 33]. In those reported in the literature it is remarkable that in 4 of 6 cases a combined replacement of the aortal and mitral valves was necessary. In 3 of the 6 patients a narrow aortic valvular ring was aditionally present and had to be corrected by an enlargement plastic procedure.

10.8 Mortality as a consequence of cardiovascular causes

In Table 12 the causes of mortality in MPS patients published in the literature are listed [3, 8, 10, 13, 20, 28, 34]. A cardiac cause was indicated in 41 % of the 114 listed patients. In addition there was a relatively high percentage of sudden and unexplained deaths, for which different explanations can be offered: an acute coronary ischaemia in preexisting coronary stenoses (see above), acute disturbances of the cardiac rhythm (see above) or acute respiratory problems.

10.9 Summary

a The patients reported in the literature are often selected and the data limited. Therefore generalized conclusions can only be drawn with great caution. Further studies are needed.

b Cardiovascular changes are very frequent in MPS, especially in MPS IH/IS, II and VI and somewhat less frequent in MPS III and IV.

c Frequently extreme coronary stenoses, caused by a thickening of the intima are found; however, they rarely cause clinical symptoms.

d Besides the coronary arteries, other arterial vessels may also be affected. In rare cases there may even be an obstruction of the abdominal aorta leading to the symptoms of aortic coarctation.

e Valvular alterations with nodular thickening, preferentially of the mitral and aortic valves, are often found leading to mild or moderate valvular regurgitation and to severe regurgitation in up to 8%. Valvular stenoses are less frequent and in most cases mild.

f Arterial hypertension is found in up to 30% of patients.

g Myocardial or endocardial alterations are not found as frequently and severely affected as the valves or coronary arteries; therefore myocardial dysfunction is relatively rare. However, there seems to be a small group of infant MPS patients presenting primarily with dilated cardiomyopathy.

Tab. 12: Causes of death in MPS patients.

Author	Reference	MPS patients	Cardiac	Pulmonary	Sudden death	Peri-OP
Krovetz	[20]	Literature	30/87	31/87	10/87	2/87
Young	[34]	MPS II	6/13	2/13	2/13	1/13
Belani	[3]	Peri-OP	2/2	0/2		2/2
	This study		2/2	0/2		0/0
	Case reports	Cardiomyopathies	7/10	0/10	3/10	-
	Total		47/114 (41%)	33/114 (29%)	15/110 (13.6%)	5/102 (5%)

h Despite the fact that the conductive system can be histologically affected, clinical arrhythmias are rare. Nevertheless there is a relatively large number of sudden deaths.
i Replacement of cardiac valves is possible.
k Among the reported deaths, cardiac causes are prominent with about 40%. In addition there is a relatively large number of unexplained deaths. These are possibly caused by arrhythmias and acute coronary ischaemias.

10.10 Conclusions

a Because even high degree valvular insufficiencies may not be detected by mere auscultation, Doppler echocardiographic investigations should be performed. In the case of valvular alterations, antibiotic endocarditis prophylaxis is recommended [1].
b Invasive procedures like valve replacement should be discussed with the parents encompassing all aspects of the disease, including mental retardation.

References

[1] American Heart Association: Commitee report on prevention of bacterial endocarditis. Circulation **70**, 1123A-1127A (1984).
[2] Beaudet, A. L., W. M. DiFerrante, G. D. Ferry, L. N. Buford, C. E. Mullins. Variation in the phenotypic expression of beta-glucoronidase deficiency. J. Pediatr. **86**, 388-394 (1975).
[3] Belani, K. G., W. Krivit, B. L. Carpenter, E. Braunlin, J. J. Buckley, J. C. Liao, T. Floyd, A. S. Leonard, C. G. Summers, S. Levine et al.: Children with mucopolysaccharidosis: perioperative care, morbidity, mortality, and new findings. J. Pediatr. Surg. **28**, 403-408 (1993).
[4] Braunlin, E. A., D. W. Hunter, W. Krivit, B. A. Burke, P. S. Hesslein, P. T. Porter, C. B. Whitley: Evaluation of coronary artery disease in the Hurler syndrome by angiography. Am. J. Cardiol. **69**, 1487-1489 (1992).
[5] Brosius, F. C., III, W. C. Roberts: Coronary Artery Disease in the Hurler Syndrome. Qualitative and quantitative analysis of the extent of coronary narrowing at necropsy in six Children. Am. J. Cardiol. **47**, 649-653 (1981).
[6] Butman, S. M., L. Karl, J. G. Copeland: Combined aortic and mitral valve replacement in an adult with Scheie's disease. Chest **96**, 209-210 (1989).
[7] Craig, W. S.: Gargoylism in a twin brother and sister. Arch. Dis. Child. **29**, 293-296 (1954).
[8] Donaldson, M. D. C., C. A. Pennock, P. J. Berry, A. W. Duncan, J. E. Cawdery, J. V. Leonard: Hurler syndrome with cardiomyopathy in infancy. J. Pediatr. **114**, 430132 (1989).
[9] Flavahan, N. A.: Atherosclerosis or lipoprotein-induced endothelial dysfunction. Circulation **85**, 19271938 (1992).
[10] Fong, L. V., S. Menahem, J. E. Wraith, C. W. Chow: Endocardial Fibroelastosis in Mucopolysaccharidosis VI. Cl. Clin. Cardiol. **10**, 362-364 (1987).
[11] Gross, D. M., J. C. Williams, C. Caprioli, B. Dominguez, R. R. Howell. Echocardiographic abnormalities in the mucopolysaccharide storage diseases. Am. J. Cardiol. **61**, 170-176 (1988).
[12] Grossmann, B. J., A. Dorfuran: In vitro comparison of the antithrombotic action of heparin and chondroitinsulfuric acid-B. Pediatrics **27**, 506-514 (1957).
[13] Hayflick, S. S. Rowe, A. Kavanaugh McHugh, J. L. Olson, D. Valle: Acute infantile cardiomyopathy as a presenting feature of mucopolysaccharidosis VI. J. Pediatr. **120**, 269-272 (1992).
[14] Herd, J. K., S. Subramanian, H. Robinson: Type III mucopolysaccharidosis: report of a case with severe mitral valve involvement. J. Pediatr. **82**, 101-104 (1973). [15] Hunter, C.: A rare disease in two brothers. Proc. Roy. See. Med. **10**, 104-108 (1917).
[16] John, R. M, D. Hunter, R. H. Swanton: Echocardiographic abnormalities in type IV mucopolysaccharidosis. Arch. Dis. Child. **65**, 746-749 (1990).
[17] Johnson, G. L., D. L. Vine, C. M. Cottrill, J. A. Noonan: Echocardiographic mitral valve deformity in the mucopolysaccharidosis. Pediatrics **67**, 401-406 (1981).
[18] Krovetz, L. J., A. E. Lorincz, G. L. Schiebler: Cardiovascular manifestations of the Hurler syndrome. Hemodynamic and angiocardiographic observations in 15 patients. Circulation **31**, 132-141 (1965).
[19] Krovetz, L. J., T. G. McLoughlin, G. L. Schiebler. Left ventricular function in children studied by increasing peripheral resistance with angiotensin. Circulation **37**, 729-737 (1968).
[20] Krovetz, L. J., G. L. Schiebler: Cardiovascular manifestations of the genetic mucopolysaccharidosis. Birth. Defects **8**, 192-196 (1972).
[21] Morrison, L. M., G. S. Bajwa, B. H. Ershoff: Prolongation of the thrombus formation time of dogs administered chondroitin sulfates A and C. Exp. Med. Surg. **28**, 188-193 (1970).
[22] Nelson, J., M. D. Shields, H. C. Mulholland: Cardiovascular studies in the mucopolysaccharidoses. J. med. Genet. **27**, 94-100 (1990).
[23] Neufeld, E. F., J. Muenzer. The Mucopolysaccharidosis. In: C. R. Scriver, A. L. Beaudet, W. S. Sly, D. Valle:

The metabolic basis of inheredited disease, Ed. 6 McGraw-Hill, New York p. 1565-1588 (1989).

[24] Renteria, V. G., Ferrans, V. J.: Intracellular collagen fibrils in cardiac valves of patients with the Hurler syndrome. Lab. Invest. **34**, 263-272 (1976).

[25] Renteria, V. G., V. J. Ferrans, W. C Roberts: The heart in the Hurler syndrome: gross, histologic and ultrastructural observations in live necropsy cases. Am. J. Cardiol. **38**, 487-501 (1977).

[26] Schieken, R. M., R. E. Kerber, V. V. Ionasescu, H. Zellweger: Cardiac manifestations of the mucopolysaccharidoses. Circulation **52**, 700-705 (1975).

[27] Spranger, J.: Inborn errors of complex carbohydrate metabolism. Amer. J. hum. Genet. 18, 489-494 (1987).

[28] Stephan, M. J., E. L. Stevens jr., R. J. Wenstrub, C. R. Greenberg, H. L. Gritter, G. F. Hodges, B. Guller. Mucopolysaccharidosis I presenting with endocardial fibroelastosis of infancy. AJDC **143**, 782-784 (1989).

[29] Tan, C. T., H. V. Schaff, F. A. Miller, Jr., W. D. Edwards, P. S. Karnes: Valvular heart disease in four patients with Maroteaux-Lamy syndrome. Circulation **85**, 188-195 (1992).

[30] Taylor, D. B., S. L. Blaser, P. E. Burrows, D. A. Stringer, J. T. Clarke, P. Thorner: Arteriopathy and coarctation of the abdominal aorta in children with mucopolysaccharidosis: imaging findings. AJR Am. J. Roentgenol. **157**, 819-823 (1991).

[31] Taylor, J., P. Thorner, D. F. Geary, R. Baumal, J. W. Balfe: Nephrotic syndrome and hypertension in two children with Hurler syndrome. J. Pediatr. **108**, 726-729 (1986).

[32] Vinallonga, X., N. Sanz, A. Balaguer, L. Miro, J. J. Ortega, J. Casaldaliga: Hypertrophic cardiomyopathy in mucopolysaccharidosis: regression after bone marrow transplantation. Pediatr. Cardiol. **13**, 107-109 (1992).

[33] Wilson, C. S., H. T. Mankin, J. R. Pluth: Aortic stenosis and mucopolysaccharidosis. Ann. Intern. Med. **92**, 496-498 (1980).

[34] Young, L D., P. S Harper: Mild form of Hunter's syndrome: clinical delineation based on 31 cases. Arch. Dis. Child. **57**, 828-836 (1982).

11 Cardiovascular involvement in Gaucher disease

E. Mengel

11.1 Introduction

Gaucher disease (GD) is the most frequent sphingolipidosis, first described by Philippe Charles Ernest Gaucher in 1882 [8]. This recessively inherited metabolic disorder is caused by deficient activity of the lysosomal enzyme, β-glucocerebrosidase. The gene encoding the lysosomal enzyme is located on the chromosome 1 (q21-31) and contains 11 exons. The phenotypic heterogeneity is explained by more than 200 recently detected mutations.

The enzyme defect leads to glucocerebroside accumulation in the lysosomes of macrophages. Lipid-laden macrophages, called Gaucher cells, are found in the organs of the reticulo-endothelial system, leading to a multisystemic disease with visceral enlargement and bone marrow displacement. According to the neurological involvement GD is subclassified into acute (type II) and chronic neuronopathic (type III) and non-neuronopathic forms (type I). The non-neuronopathic form, which represents >90 % of cases, is characterised by hepatosplenomegaly, anaemia, thrombocytopenia and bone involvement. Lung involvement is seen in 10-20% of patients. Other organs like kidney, heart and intestinal bowel are rarely involved. Acute neuronopathic GD refers to progressive neurodegeneration including bulbar involvement in the first year of life. In chronic neuronopathic GD the systemic disease is usually more aggressive with onset in the first or second year of life. Neurological progression is less dramatic and more variable than in the acute neuronopathic form. Poor psychometric function and oculomotor abnormalities are often the first signs of neurological involvement which may progress to ataxia, seizures, spasticity, dementia and death.

Enzyme replacement therapy (ERT) is the therapy of choice in non-neuronopathic patients. It is very effective in reversing all haematological and visceral changes [2]. Bone involvement requires doses of 50-60 U / kg body weight twice monthly [12]. Also chronic neuronopathic patients may benefit from ERT: neurological stabilisation or improvement is seen in some patients on very high doses (100-120 U / kg body weight two times per month. In acute neuronopathic patients resulting from prenatal neuronal damage, ERT is ineffective and may prolong pain and suffering [19].

Cardiac involvement seems to be a rare but serious complication in GD. In anecdotal case reports, constrictive pericarditis and interstitial infiltration of the myocardium by Gaucher cells causing restrictive cardiomyopathy were described in non-neuropathic patients. Moreover, in 1995 Abrahamov et al. [1] and Chabas et al. [5] characterised genetically a variant of GD with predominant progressive calcification of heart valves and mild neurological involvement. These patients are homozygous for the point mutation D409H (1342C). Finally, severe pulmonary hypertension without radiological lung involvement is a serious cardiovascular complication in GD. Mostly splenectomised women seem to be affected.

11.2 Heart

11.2.1 Pericarditis

Pericarditis has been reported in a limited number of GD patients [3, 11, 14]. All authors suggest that intrapericardial haemorrhage caused by haemorrhagic diathesis, which is commonly seen in GD, attributes to this complication. In one patient, severe pericardial haemorrhage with subsequent cardial tamponade was described. Tamari et al. [14] speculated that minor trauma, i.e. coughing, caused intra-pericardial bleeding. Constriction and calcification occurred later in life and might be one of the final manifestations. Histologically, pericardial infiltration of Gaucher cells was not found. In all these cases, severe, overwhelming multisystemic disease, other non-reticulo-endothelial organ involvement and poor outcome were evident.

11.2.2 Myocardial infiltration

Left ventricular dysfunction is rare in GD [6, 13]. However, systematic evaluation with echocardiography and ECG in larger GD cohorts is lacking. Cardiac biopsies have only been taken in a few cases. In these patients, interstitial infiltration of the myocardium with Gaucher cells was present. In the case report of Smith et al. [13], mainly the left ventricle was involved; in comparison the right ventricle was only mildly affected. No evidence of myocarditis was present at the time of biopsy.

Whether myocardial infiltration causes fatal left ventricle dysfunction even when other severe complications, i.e. sepsis or pulmonal or portal hypertension, occur has not been investigated. In our group of 70 patients, we have observed sign of left ventricle dysfunction and echointense myocardium in 2 women on low dose, unsuccessful ERT and more than 15 years after splenectomy. Biopsies were not taken. Under treatment with higher doses of imiglucerase, the patients remained stable for several years.

11.2.3 Heart valve calcification

Casta et al. [4] initially reported aortic and valve calcification in a patient with GD. Ten years later, Japanese authors described heart valve calcifications in GD patients with mild neurological involvement and moderate systemic disease [16]. They also observed further unusual abnormalities like corneal opacities, hydrocephalus, deafness, deformed toes and leptomeningeal fibrous thickening in these patients. In 1995, Chabas et al. and Abrahamov found independently in these patients homozygosity for the mutation D409H (1342C) [1, 5]. Thereafter, the Japanese patients were also found to be homozygous for this mutation [17]. Recently, unpublished information as received indicates that in several Mediterranean countries (Egypt, Greece, Turkey) Gaucher patients with heart valve calcifications and the homozygous mutation D409H/D409H are known.

The presence of Gaucher cells in the valve tissue was documented by Veinot et al. [18]. They suggested that Gaucher cells and lymphocytes regulate the process of calcification by releasing matrix proteins and cytokines. This pathology may also explain calcifications in other tissues. However, it remains unclear why only patients with this specific mutation develop heart valve calcifications.

11.3 Cor pulmonale

It was nearly four decades ago that the first report of fatal pulmonary hypertension causing heart failure in non-neuronopathic GD appeared. Since then about a dozen cases have been reported in the literature. In most cases the lung was not infiltrated with Gaucher cells. The pathophysiology is not well understood, but it seems reasonable to assume that macrophage dysfunction – macrophages are the primary storage cells – causes pulmonary vascular alteration [11, 13, 15].

Harats et al. [9] and Elstein et al. [7] were afraid that ERT could play a role in pulmonary hypertension in GD because they observed pulmonary hypertension in treated patients. Recently Mistry et al. [10] consecutively investigated 134 treated and non-treated patients by Doppler echography. They found no correlation with ERT. Moreover, they identified risk factors for pulmonary hypertension. Like in other patients with pulmonary hypertension, asplenia and female sex were found in the majority of the patients. Additional no or unsuccessful ERT, positive family history, ACE I allele and more severe glucocerebrosidase gene mutations were significant risk factors [10]. These findings suggest that different genetic and epigenetic determinants influence the appearance of pulmonary hypertension in GD. The observations of Mistry et al. [10] emphasised by our own unpublished experience show that the outcome of patients with GD and severe pulmonary hypertension treated by a combination of ERT, vasodilators and coumadin is much better than expected in comparison to patients with pulmonary hypertension alone.

11.4 Conclusion

Cardiovascular involvement and consecutive heart failure are rare but serious findings in GD. Careful and regular examinations including echography are warranted for all patients, especially in those with the homozygous mutation D409H and in splenectomised female patients. Risk factors and pathophysiological explanations are described above. However, in the area of ERT heart complications should be prevented and successfully treated.

References

[1] Abrahamov, A., D. Elstein et al.: Gaucher's disease variant characterised by progressive calcification of heart valves and unique genotype. Lancet **346** (8981), 1000-1003 (1995).

[2] Barton, N., R. Brady et al.: Replacement therapy for inherited enzyme deficiency - macrophage-target glucoceribrosidase for Gaucher's disease. New Engl. J. Med. **324**, 1679-1682 (1991).

[3] Benbassat, J., H. Bassan et al.: Constrictive pericarditis in Gaucher's disease. Am. J. Med. **44**(4), 647-652 (1968).

[4] Casta, A., K. Hayden et al.: Calcification of the ascending aorta and aortic and mitral valves in Gaucher's disease. Am. J. Cardiol. **54**(10), 1390-1391 (1984).

[5] Chabas, A., B. Cormand et al.: Unusual expression of Gaucher's disease: cardiovascular calcifications in three sibs homozygous for the D409H mutation. J. Med. Genet. **32**(9), 740-742 (1995).

[6] Edwards, W. D., H. P. Hurdey, 3rd et al.: Cardiac involvement by Gaucher's disease documented by right ventricular endomyocardial biopsy. Am. J. Cardiol. **52**(5), 654 (1983).

[7] Elstein, D., M. W. Klutstein et al.: Echocardiographic assessment of pulmonary hypertension in Gaucher's disease. Lancet **351**(9115), 1544-1546 (1998).

[8] Gaucher, P.: De l'epithelimoma primitif de la rate. Faculte de Medecine. Paris (1882).

[9] Harats, D., R. Pauzner et al.: Pulmonary hypertension in two patients with type I Gaucher disease while on alglucerase therapy. Acta Haematol **98**(1), 47-50 (1997).

[10] Mistry, P., S. Sirrs et al.: Pulmonary hypertension in type 1 Gaucher's disease: genetic and epigenetic determinants of phenotype and response to therapy. Mol. Genet. Metab. **77**(1-2), 91 (2002)

[11] Roberts, W. C., D. S. Fredrickson: Gaucher's disease of the lung causing severe pulmonary hypertension with associated acute recurrent pericarditis. Circulation **35**(4), 783-789 (1967).

[12] Rosenthal, D. I., S. H. Doppelt et al.: Enzyme replacement therapy for Gaucher disease: skeletal responses to macrophage-targeted glucocerebrosidase. Pediatrics **96** (4 Pt 1), 629-637 (1995).

[13] Smith, R. L., G. M. Hutchins, et al.: Unusual cardiac, renal and pulmonary involvement in Gaucher's disease. Intersitial glucocerebroside accumulation, pulmonary hypertension and fatal bone marrow embolization. Am. J. Med. **65**(2), 352-360 (1978).

[14] Tamari, I., M. Motro et al.: Unusual pericardial calcification in Gaucher's disease. Arch. Intern. Med. **143**(10), 2010-2011 (1983).

[15] Theise, N. D., P. C. Ursell: Pulmonary hypertension and Gaucher's disease: logical association or mere coincidence? Am. J. Pediatr. Hematol. Oncol. **12**(1), 74-76 (1990).

[16] Uyama, E., K. Takahashi et al.: Hydrocephalus, corneal opacities, deafness, valvular heart disease, deformed toes and leptomeningeal fibrous thickening in adult siblings: a new syndrome associated with -glucocerebrosidase deficiency and a mosaic population of storage cells. Acta. Neurol. Scand. **86**(4), 407-420 (1992).

[17] Uyama, E., M. Uchino et al.: D409H/D409H genotype in Gaucher-like disease. J. Med. Genet. **34**(2), 175 (1997).

[18] Veinot, J. P., D. Elstein et al.: Gaucher's disease with valve calcification: possible role of Gaucher cells, bone matrix proteins and integrins. Can. J. Cardiol. **15**(2), 211-216 (1999).

[19] Vellodi, A., B. Bembi et al.: Management of neuronopathic Gaucher disease: a European consensus. J. Inherit. Metab. Dis. **24**(3), 319-327 (2001).

12 Selenium deficiency in children with cystic fibrosis and phenylketonuria-metabolic and echocardiographic findings during sodium selenite therapy

E. Kauf, L. Vogt, J. Seidel, K. Winnefeld, H. Richter, H. Vogel, H. Dawczynski, A. Forberger, D. Schlenvoigt

A sufficient nutritional supply of the essential trace element selenium is necessary to maintain the function of selenoenzymes. Glutathione peroxidase (GSH-Px), an important enzyme in the antioxidative system, and type I iodothyronine-5' deiodinase, which catalyses the hepatic conversion of thyroxine to triiodothyronine, are known selenoenzymes [1, 2, 18]. Selenium deficiency in man produces clinical symptoms of Keshan disease (endemic cardiomyopathy), Kashin-Beck disease (endemic osteoarthropathy) and congestive cardiomyopathy [11, 16, 22]. Patients with cystic fibrosis (CF) often develop selenium deficiency due to disturbed enteral digestion and absorption (steatorrhoea) [5, 10, 14, 15, 20] whereas phenylketonuria (PKU) patients always suffer from selenium deficiency if a phenylalanine restricted diet without selenium supplementation is used over a long period of time [3, 8, 12, 13]. We investigated a group of CF patients and a group of PKU patients with respect to selenium deficiency, metabolic consequences and cardiac function.

12.1 Patients and methods

Two patient groups, CF (n = 32, age range 13.8 ± 7.4 years; 17 male, 15 female) and PKU (n = 17, age range 8.21 ± 3.74 years; 9 male, 8 female) were investigated prior to and during 3 months of sodium selenite therapy (115 µg selenium/m^2/body surface area/day). Fasting plasma and blood cell selenium (B-Se), GSH-Px, total thyroxine (T4), triiodothyronine (T3), reverse triiodothyronine (rT3), total cholesterol (t-chol), HDL-cholesterol (HDL) and LDL-cholesterol (LDL) were determined. Selenium was analysed using the AAS3/Hg hydride system [6]. Plasma GSH-Px determinations were done according to Paglia [17]. T4, T3, rT3, fT4 and fT3 were determined by commercial radioimmunoassays (Henning, Serono) and plasma lipids by test kits from Boehringer-Mannheim. LDL was calculated using the Friedewald formula. Echocardiographic investigations were always performed by the same operator using the Sonoline SL-2 (Siemens) with 3.5 MHZ transducer and Doppler ultrasound system. Left ventricular cardiac index (LCVI) results (1/min/m² body surface area) were calculated by Wessel's formula [21]. The diastolic function of the left ventricle was determined by the ratio of early diastolic inflow to late diastolic flow (Ve/Va) at the mitral valve.

For statistical analysis the arithmetic means of determined values before and after selenium therapy were compared by paired t-test, correlations were calculated by Pearson's coefficients.

12.2 Results and discussion

Prior to selenium therapy, plasma Se levels were diminished in both patient groups but normalized after 3 months of therapy (Table 1). The results of the blood Se contents demonstrated a slight Se deficiency in the CF group whereas that of the PKU group was more severe. Connected with this were the diminished levels of plasma GSH-Px which originates mainly in the liver [9]. Despite existing Se deficiency, the absolute val-

Tab. 1: Blood cell selenium (B Se), selenium (P Se), glutathione peroxidase (P GSH-Px), thyroid hormones (T4, T3, rT3) and lipids (t-Chol, LDL) in the plasma of CF and PKU patients prior to and during oral sodium selenite therapy.

		Before	Signif.[1]	After	Normal Range
CF group (n = 32)		$\bar{x} \pm si$		$\bar{x} \pm s$	
§P Se	µmol/l	0.69 ± 0.21	**	0.96 ± 0.27	1.03 ± 0.20
§B Se	µmol/l	1.14 ± 0.40	ns	1.16 ± 0.34	1.28 ± 0.22
§P GSH-Px	U/l	86.3 ± 27.3	**	111.7 ± 32.5	123 ± 27
§T$_4$	nmol/l	102.3 ± 27.4	ns	98.5 ± 28.9	58 – 154
§T$_3$	nmol/l	2.14 ± 0.56	*	2.39 ± 0.55	1.2 – 2.8
§rT$_3$	ng/ml	0.18 ± 0.04	ns	0.17 ± 0.04	0.1 – 0.35
§T$_4$/T$_3$		49.8 ± 14.2	**	43.1 ± 12.8	
§t-Chol	mmol/l	3.39 ± 1.28	ns	3.18 ± 0.76	2.8 – 5.0
§LDL	mmol/l	2.16 ± 1.04	*	1.73 ± 0.53	1.4 – 3.6
PKU group (n = 17)					
§P Se	µmol/l	0.17 ± 0.06	***	1.08 ± 0.20	
§B Se	µmol/l	0.38 ± 0.23	***	1.07 ± 0.21	
§P GSH-Px	U/l	34.5 ± 21.9	***	123.3 ± 25.4	
§T4	nmol/l	125.2 ± 14.3	***	103.0 ± 22.1	
§fT4	pmol/l	31.1 ± 3.90	***	25.3 ± 4.40	10 – 25
§T3	nmol/l	2.65 ± 0.42	ns	2.43 ± 0.41	
§fT3	pmol/l	8.18 ± 1.47	ns	8.65 ± 1.03	3.4 – 8.5
§rT3	ng/ml	0.25 ± 0.05	***	0.15 ± 0.05	0.1 – 0.35
§T4/T3		48.4 ± 10.8	ns	43.5 ± 12.4	
§fT4/fT3		3.90 ± 0.82	***	2.98 ± 0.76	
§t-Chol	mmol/l	4.47 ± 0.64	**	3.84 ± 0.59	
§LDL	mmol/l	2.65 ± 0.59	**	2.14 ± 0.55	

1 paired t-test;   * $p \leq 0.05$;   ** $P \leq 0.01$;   *** $P \leq 0.001$

ues of lipids and thyroid hormones in both groups before and during sodium selenite therapy were, with exception of initial elevated fT4 in PKU patients, normal (Table 1). The left cardiac function (LCVI) was reduced in both groups (Table 2). The mean LCVI in CF patients with a less severe Se deficiency was smaller than in PKU patients. This may be interpreted as being related to an often existing vitamin E deficiency in CF patients [19] (our patients: 4.31 ± 2.69 µg/ml; normal > 4.7). Vitamin E deficiency is rare in PKU patients.

The dependence of the diminished LVCI on Se deficiency is demonstrated by the significant enhancement of this parameter during Se substitution in both patient groups. The major mechanism of myocardial disturbance in Se deficiency could be an increase in free radicals as the Se-dependent GSH-Px is low. The PKU patients with a more pronounced Se deficiency showed the greater enhancement of LCVI, moreover, there was a significant enhancement of the diastolic function of the left ventricle. In the CF group the enhancement of LVCI under Se therapy was also significant, but without improvement in diastolic function.

As type I iodothyronine-5'-deiodinase is Se dependent [1, 2], there were significant changes in thyroid hormones during therapy of the severe Se deficiency in the PKU group. The significant decrease in T4 was a sign of improved T4 conversion, but no parallel T3 enhancement in plasma was observed. Nevertheless, the reduced T4/T3 ratio, fT4/fT3 ratio and diminished rT3 levels showed increased efficacy of T4 to T3 conversion by Se in our patients. Similar results have been reported in Se-deficient rat pups [4] and in other PKU children who received only 2 weeks of Se supplementation (1 µg/kg/day) [7]. The corresponding influences in the CF group were less dramatic with an insignificant T4 reduction, a T3 elevation and a reduction in T4/T3 ratio. In addition to the effect of enhanced GSH-Px function during Se replacement therapy, the improved thyroid hormone metabolism certainly plays a role in cardiac function (oxidative phosphorylation). In the PKU group, we found a significant correlation between Se increase in blood cells and

Tab. 2: Echocardiographic left ventricular cardiac index (LVCI) and diastolic flow ratio ($V_{E/A}$) at the mitral valve in CF and PKU patients prior to and during 3 months of sodium selenite therapy.

		Before	Signif.	After	Normal Range
CF group (n = 32)		$\bar{x} \pm s$		$\bar{x} \pm s$	
§ LVC1 § BSA	1/min/m²	2.38 ± 0.73	**	2.83 ± 0.72	2.50 – 4.50
§ $V_{E/A}$		1.58 ± 0.37	ns	1.47 ± 0.39	1.4 – 2.0
PKU group (n =17)					
§ LVC1 § BSA	1/min/m²	2.67 ± 0.73	*	3.36 ± 0.81	
§ $V_{E/A}$		1.70 ± 0.35	*	1.91 ± 0.45	

¹paired t-test; * $p \leq 0.05$; ** $p \leq 0.01$; *** $p \leq 0.001$

LM during therapy (r = 0.83) and a correlation between Se increase in blood cells and decrease in plasma rT3 (r = 0.69). Other indicators of Se replacement effect are total cholesterol and LDL-cholesterol levels in plasma. The hepatic metabolism of cholesterol is mainly influenced by thyroid hormones, in particular T3. Increased cholesterol and LDL-cholesterol levels as well as disturbed cardiac function (latent cardiac insufficiency) are common in hypothyroid patients. The significant drop in total cholesterol and LDL-cholesterol in our PKU patients and the similar but smaller reaction in CF patients under Se therapy demonstrated indirectly the relation between Se and thyroid hormone metabolism in hepatic cells. Long-term reduction of plasma cholesterol by sufficient Se supply has probably a beneficial effect on cardiac function. Other unknown Se-dependent factors like selenoproteins [9] and the partly Se-influenced synthesis of prostaglandins may support an effect of Se on myocardial function. The following can be concluded from our results:
- in CF a combined deficiency of Se and vitamin E often exists.
- PKU patients under long-term dietary control without Se supplementations always develop a severe Se deficiency.
- Se deficiency in children creates a reduced cardiac function with cardiomyopathy.
- oral sodium selenite therapy abolishes Se deficiency with restitution of GSH-Px activity and type I iodo-thyronine-5'-deiodinase with subsequent increase in cardiac output.
- in all patients with unclear reduced cardiac function, Se deficiency should be excluded.

References

[1] Beckett, G J., D. A. MacDougall, F. Nicol, J. R. Arthur: Inhibition of type I and type II iodothyronine deiodinase activity in rat liver, kidney and brain produced by selenium deficiency. Biochem. J. **259**, 887-892 (1989).

[2] Behne, D., A. Kyriakopulous, H. Meinold, J. Köhrle: Identification of type I iodothyronine 5'α-deiodinase as a selenoenzyme. Biochem. Biophys. Res. commun. **173**, 1143-1149 (1990).

[3] Calomme, M.: Supplementation of phenylketonuria children with a bacterial selenosource. 30th Annual Meeting of society for the study of inborn errors of metabolism, Leuven Abstr. P14 (1992).

[4] Chanoine, P., S. Alex, S. Stone, S. L. Fang, J. L. Leonhard, L. E. Brotherman: Normal serum T3 postnatal source and increased reverse T3 levels in selenium-deficient rat pups. Pediatr. Res. **33**, 92, Abstract 529 (1993).

[5] Chase, H P., M. A. Long, M. H. Lavin: Cystic fibrosis and malnutrition. J. Paediatr. **95**, 337-347 (1979).

[6] Dawczynski, H. K. Winnefeld, E. Tennigkeit: Die quantitative Bestimmung des Selen im Serum und Vollblut mit dem Gerätesystem Atom-Absorptions-Spektralphotometer AAS 3/Hg/Hydridsystem HS. Carl-Zeiss-Applikationsinformation - optische Analysenmeßgeräte **5**, 1-12 (1986).

[7] Francois, M., J. van Caillie-Bertrand, J. Vandeerpas, G. Massa, M. Calomme, D. van den Berghe: Effect of selenium deficiency and supplementation on serum levels of thyroid hormones of children with phenylketonuria (PKU). Pediatr. Res. **33**, 93 Abstr. 537 (1993).

[8] Greeves, G. D. J. Carson, B. G. Craig, D. McMaster. Potentially life-threatening cardiac dysrhythmia in a child with selenium deficiency and phenylketonuria. Acta Paed. Scand. **79**, 1255-1262 (1990).

[9] Haas, H. J.: Selenoproteine in Bakterien, Hefen, Tieren und Menschen. In: Mengen- und Spurenelemente, 13. Arbeitstagung Friedrich-Schiller-Universität. M. Anke, H. Bergmann, R. Bitsch, W. Dorn, G. Flachowsky B. Groppel, H. Gürtler, I. Lombeck, B. Lukkas, H. J. Schneider (Hrsg.). Verlag MTV Hammerschmidt GmbH, Gersdorf, S. 438 (1993).

[10] Kauf, E., H. Dawczynski, L. Vogt, H. Hofman, P. Möckel. Selen und Vitamin E im Blut. Aspekte der chronischen Probleme von Mukoviszidosepatienten. Zentr. Bl. Pharm. Lab. Diag. **127**, 400-402 (1988).

[11] Keyou, G., G. Yang: The epidemiology of selenium deficiency in the etiological study of endemic diseases in China. Amer. J. Clin. Nutr. Suppl. **57**, 259S-263S (1993).

[12] Lombeck, I. K. Kasparek, D. Bachmann, L. E. Feinendegen, H. J. Bremer: Selenium requirements in patients with inborn errors of amino acid metabolism and selenium deficiency. Eur. J. Pediatr. **138**, 65-68 (1980).

[13] Lombeck, I. K. Kasparek, H. D. Harbisch, K. Bekker, E. Schumann, W. Schröter, L. E. Feinendegen, H. J. Bremer: II. Selenium content of serum, whole blood, hair and the activity of erythrocyte glutathione peroxidase in dietetically treated patients with phenylketonuria and maple-syrup-urine disease. Eur. J. Pediatr. 128, 213-223 (1978).

[14] Lloyd-Still, J. D., H. E. Ganther: Selen and glutathion peroxidase levels in cystic fibrosis. Pediatrics **65**, 1010-1012 (1980).

[15] Neve, J., R. van Geffel, M. Hanocq, L. Molle: Plasma and erythrocyte zinc, copper and selenium in cystic fibrosis. Acta Paediatr. Scand. **72**, 437-440 (1983).

[16] Oster, O., W. Prellwitz, W. Kasper, T. Meinertz: Congestive cardiomyopathy and the selenium content of serum. Clin. chim. Acta **128**, 125-132 (1983).

[17] Paglia, D. E., W. N. Valentine: Studies on the quantitative and qualitative characterization of erythrocyte glutathione peroxidase. J. Lab. Clin. Med. **70**, 158-168 (1967).

[18] Rotruck J. T., A. L. Pope, H. E. Ganther, W. G. Hoekstra: Prevention of oxidative damage to rat erythrocytes by dietary selenium. J. Nutr. **102**, 689-692 (1972).

[19] Skopnik, H. M. Karl, G. Kusenbach, U. Bergt, M. Geron, G. Heimann: Einfluß des Ernährungsstatus auf die Absorptionskinetik von Vitamin E bei Mukoviszidose. Klin. Päd. **202**, 43-49 (1990).

[20] Van Hubbard, S. G. Barbero, H. P. Chase: Selenium and cystic fibrosis. J. Pediatr. **96**, 421 (1980).

[21] Wessel, A., D. G. W. Onnasch, M. P. Heintzen, W. Berdau, P. H. Heintzen: Sektorechokardiographische Untersuchungen zum normalen Wachstum des rechten und linken Ventrikels im Kindesalter. In: R. Erbel, J. Meyer, R. Brennecke (Hrsg.): Fortschritte der Echokardiographie; Springer Verlag, Berlin, Heidelberg, New York, Tokyo. S. 78-87 (1985).

[22] Yang, G. Q.: Keshan disease. In: an endemic selenium-related deficiency disease. R. K. Chandra (Hrsg.): Trace elements in nutrition of children. Nestlé nutrition workshop series. Vol. 8, Raven Press, New York. S. 273-287 (1985).

13 Fabry disease – a progressive multi-systemic lysomal storage disorder

M. Beck

13.1 Pathophysiology

Glycosphingolipids are essential constituents of cell membranes and are widely distributed in human tissues. They consist of a hydrophilic complex carbohydrate chain and a hydrophobic ceramide. In ceramide, an amino alcohol (sphingosine) is acylated with a long-chain fatty acid through an amide linkage (Fig. 1). Many lysosomal enzymes are involved in the stepwise degradation of these sphingolipids. A genetic defect of one of these enzymes leads to disorders that are characterized by progressive storage in affected organs and functional impairment. In Fabry disease, glycosphingolipids with terminal α-galacto-

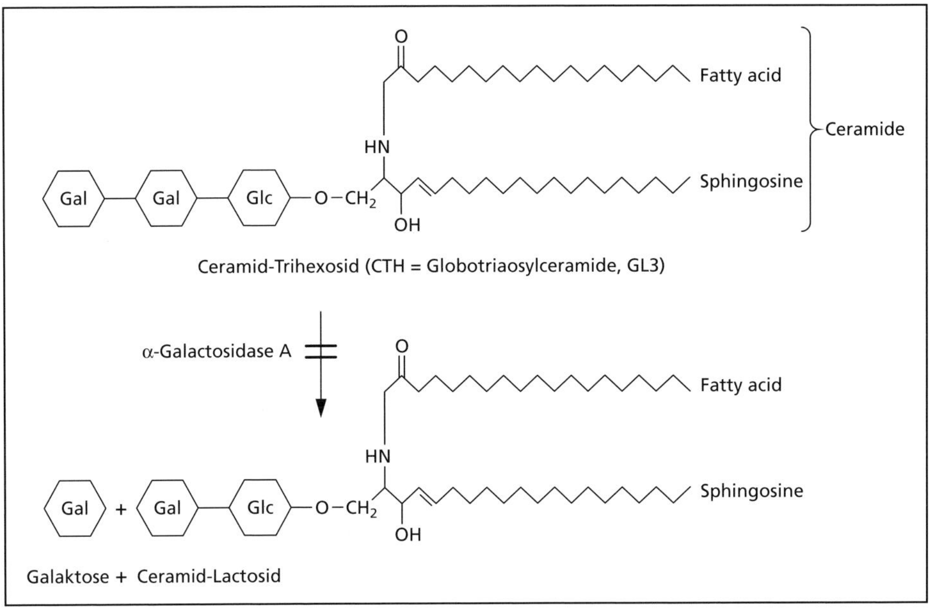

Fig. 1: Degradation of globotriaosylceramide by the lysosomal enzyme α-galactosidase A. This enzyme is deficient in Fabry disease.

syl moieties, predominantly globotriaosylceramide (= Gb3) and, to a lesser extent digalactosylceramide and blood group substances, are accumulated in a variety of cell types due to the genetic defect of the lysosomal enzyme α-galactosidase A [1]. Progressive deposition of globotriaosylceramide is found in endothelial cells, smooth muscle cells of blood vessels, ganglion cells and many cell types of the kidneys, heart and eyes. In contrast, digalactosylceramide has been detected only in the right heart and the lungs. The accumulation of storage material in a broad variety of cell types leads to a multisystemic disorder characterized by acroparesthesias, skin and eye abnormalities, cardiomyopathy, kidney dysfunction and cerebrovascular complications.

The enzyme deficient in Fabry disease, α-galactosidase A (α-D-galactoside galactohydrolase, E.C. 3.2.1.22), is a glycoprotein of approximately 101 kDa and has a homodimeric structure (Fig. 2) [2]. It contains 5 - 15 % asparagine-linked complex and high-mannose oligosaccharide chains. It must be discriminated from the enzyme α-galactosidase B (= α-N-acetylgalactosaminidase, EC 3.2.1.49), which is genetically distinct and has different catalytic properties [3]. Whereas α-galactosidase A splits galactose from substrates containing terminal α-galactosidic residues, α-galactosidase B is responsible for the degradation of substrates with α-acetylgalactosaminyl moieties. The deficiency of α-galactosidase B leads to the lysosomal storage disorder known as Schindler disease.

13.2 Symptoms

13.2.1 Pain

In males (and females) affected by Fabry disease, first symptoms arise during childhood: boys and girls complain of constant burning pain (acroparesthesias) and tingling, especially in the toes and fingers. In addition, they experience episodic crises of neuropathic pain in the hands and feet that are often triggered by fever, exercise or emotional stress (Fig. 3). Also abdominal pain has been reported that may mimic appendicitis or renal colic.

In order to characterize the physiological abnormality that causes acroparesthesias, Luciano and co-workers performed systematic investigations in 22 patients with Fabry disease [4]. In most patients, nerve conduction was normal except for an increased frequency of median nerve entrapment at the wrist in 6 patients. In 19 of 20 patients, sympa-

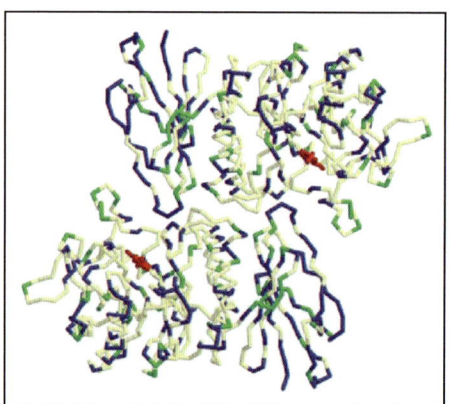

Fig 2: Three-dimensional structure of alpha-galactosidase. (Garman, S. C., L. Hannick, A. Zhu, D. N. Garboczi: the 1.9-α-structure of alpha-N-acetylgalactosaminidase: molecular basis of glycoside deficiency diseases. Structure 10, 425-434 (2002). Reproduced with permission of the authors).

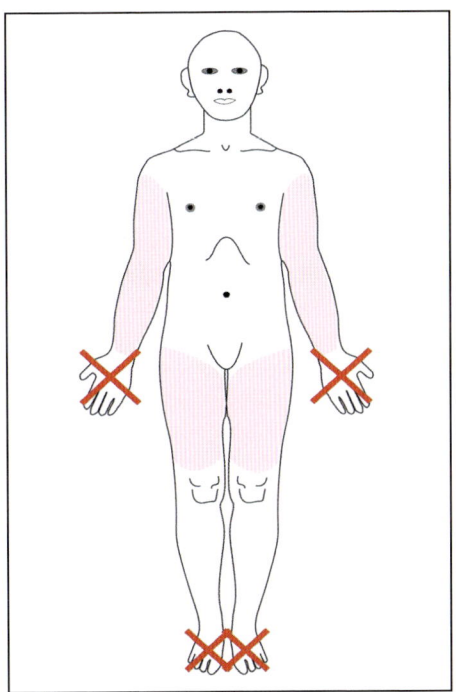

Fig. 3: Area of the body affected by acroparaesthesias. Pain begins in the hands and feet and radiates to the shaded areas.

thetic skin responses were preserved. The quantitative sensory testing (QST) showed increased or immeasurable cold and warm detection thresholds, whereas cold thresholds were more often abnormal than warm thresholds. From their findings, the authors suggest that the neuropathy of Fabry disease is characterized by an increased prevalence of median nerve entrapment at the wrist and by thermal afferent fibre dysfunction in a length-dependent fashion, with greater impairment of cold than warm sensation.

Histological examination of intradermal nerve fibres revealed signs of lipid accumulation and degeneration of small unmyelinated nerve fibres, however, lipid inclusions were not found in Schwann cells. Those cells probably serve as a metabolic barrier that protects the larger myelinated fibres, while the smaller unmyelinated fibres are more vulnerable to lipid infiltration. In addition to small fibre involvement, accumulation of storage material in cells of the dorsal root (sensory) ganglia may also play a role in the pathogenesis of neuropathic pain.

13.2.2 Skin

Already in childhood, punctate dark red to blue-black angiectases (angiokeratoma) develop in the skin. These lesions are often flat or slightly raised and do not blanch with pressure. Angiokeratoma increase in number and size with age and are most commonly seen on the hips, back, thighs, buttocks, penis and scrotum. Also the oral mucosa and conjunctiva may be involved.

13.2.3 Autonomic dysfunction

A very common symptom seen in affected males and females is hypohidrosis (reduced sweating) or even anhidrosis (inability to sweat) which may be explained by autonomic nervous system dysfunction and also by the accumulation of globotriaosylceramide in the eccrine cells. In addition, tear and saliva formation is often reduced. As a consequence of disturbed autonomic nervous function, abnormalities of cerebrovascular reactivity, cardiac rhythm and gastrointestinal complaints are observed. Further gastrointestinal symptoms include postprandial pain, nausea and diarrhoea. Even diverticulosis with perforation has been described [5].

13.2.4 Eyes

In the eyes, corneal opacities are seen on slit-lamp examination and appear as whorled streaks extending from a central vortex to the periphery of the cornea. Those changes are called "cornea verticillata" (Fig. 4). Anterior, capsular or subcapsular and posterior lenticular deposits may also be detected. These ocular changes do not impair vision, but later in the illness, retinal complications, hypertension and uraemia may occur.

13.2.5 Heart

After the age of 30 years, the major morbid symptoms of the disease result from the progressive glycosphingolipid deposition in the cardiovascular and renal systems. Heart manifestation includes left ventricular enlargement, valvular abnormalities and arrhythmias. Details of cardiac involvement are given in the chapter by Kampmann.

13.2.6 Kidney

In the kidney there is a mixed glomerular and tubular disease marked by proteinuria, haematuria and tubular dysfunction. Often already in teenage years, Gb3 deposition begins in the glomerulus and is associated with mild proteinuria and occasionally microhaematuria. On light microscopy, glomeruli show hypertrophic podocytes, distended with foamy appearing vacuoles, mesangial widening and varying degrees of glomerular obsolescence. In later stages, proximal tubules and interstitial cells may show lipid accumulation leading to urinary concentration defects. Urine sediment Gb3 is elevated secondary to the presence of sloughed, lipid-laden, renal tubular epithelial cells. On polarization microscopy, birefringent lipid globules with characteristic "maltese crosses" can be detected free in the urine and in desquamated tubular epithelial cells.

Chronic renal insufficiency may begin at the second decade of life [6]. Deposition in capillary, arterial and arteriolar endothelial cells are signs of vascular involvement. Later, glomerular fibrosis and atrophy, interstitial fibrosis, and tubular atrophy take place leading to gradual deterioration of renal function and finally to end-stage renal failure. In a survival study conducted at the NIH, Branton and co-workers found that 50% of patients developed end-stage renal disease by 47 years, with a range of 21 to 56 years [6].

13.2.7 Cerebrovascular system

In Fabry disease, there are several factors that contribute to cerebral complications: glycosphingolipid deposition has been found in the wall of the small arteries and arterioles and ectasia of intracranial arteries due to involvement of the vascular smooth muscle. The lesions affect predominantly the vertebro-basilar region, the reason for this

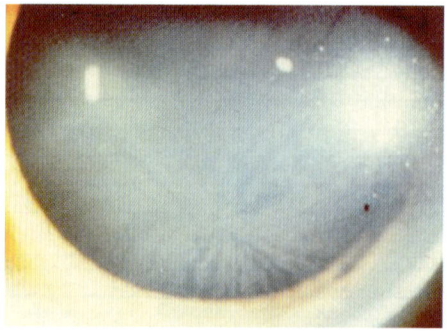

Fig. 4: Corneal opacity (cornea verticillata) observed by slit-lamp examination

distribution being unclear. In addition, patients with Fabry disease have a higher incidence of both venous and arterial intravascular thrombosis. Finally, in affected individuals an increased resting cerebral blood flow has been reported [7]. Studies performed on skin and brain specimens revealed enhanced nitrotyrosine staining in dermal and cerebral blood vessels suggesting a chronic alteration in the nitric oxide pathway [8]. MRI studies reveal white matter hyperintensities or basal ganglia infarcts [9], alterations that may lead initially to a diagnosis of multiple sclerosis. Clinical manifestations of the cerebrovascular system include head pain, vertigo, nystagmus, nausea with vomiting, hemiparesis, diplopia, and gait ataxia. In general, epileptic seizures do not occur in patients with Fabry disease, however, cognitive deficits, personality changes and psychotic behaviour are not uncommon. There is a high risk of chronic depression in both males and females that may lead to alcoholism and suicide. as hemizygotes, although to a more variable degree [11]. MacDermot and coworkers found that in males with Fabry disease, the median cumulative survival was 50 years, which represents an approximately 20 year reduction of life span [12]. In affected females life span was reduced to 70 years, representing a reduction of about 15 years compared to the general population [13]. The variable manifestation seen in heterozygotes can be explained by the Lyon hypothesis that predicts that in X-linked diseases, carriers are mosaics of normal and mutant cells in varying proportions. This assumption has been substantiated by the observation of female monozygotic twins who showed different phenotypic expression due to uneven X-inactivation [14].

The incidence of Fabry disease has been calculated to be 0.21 per 100 000 live births (0.42 per 100 000 male live births) in the Netherlands [15] and 0.85 cases per 100 000 live births in Australia [16].

13.3 Genetics

The gene encoding α-galactosidase A maps to Xq22.11 and has a size of 12-kb. It contains seven exons varying in length from 92 to 291 bp. To date, more than two hundred mutations have been detected occurring randomly throughout the gene. Mutations that have been observed in affected families include large rearrangements, insertions, deletions and point mutations [10]. Total gene deletions have not been identified.

From recent clinical studies it has become clear that in Fabry disease females are affected in the same manner

13.4 Treatment

Until recently, only supportive therapy was available for individuals affected by Fabry disease. Pain management has been shown to be very difficult, and there is little relief for many patients although they use a large number of analgesics. Sodium-blocking agents such as carbamezepine and/or phenytoin are commonly used; however, these drugs have also been shown to be of poor therapeutic value in Fabry disease, and to treat pain effectively, often therapeutic levels are required far above those generally used for treating epilepsy [17]. The episodes of pain may be treated by centrally acting narcotic analgesics such as morphine. Gabapentin, a recently introduced anti-epileptic drug used as an adjuvant in partial and secondarily generalised tonic-clonic seizures, has been proven rather effective in neuropathic pain, such as diabetic neuropathy and postherpetic neuralgia [18].

The hypohidosis (anhidrosis) can be managed by minimizing intense physical exertion and having the patient work in a cool environment. Angiokeratomas can be removed by argon laser therapy if desired for cosmetic reasons.

As already pointed out, renal insufficiency is the most frequent late complication in patients with Fabry disease. As soon as proteinuria and renal impairment develop, antihypertensive therapy must be initiated which may also be necessary for treating cardiovascular complications. Angiotensin antagonists are widely recommended. When end-stage renal failure occurs, chronic haemodialysis becomes necessary. Given the poor survival on dialysis, alternative strategies must be considered to manage renal dysfunction. Kidney transplantation has been performed in a large number of patients, and it could be shown that this procedure leads to correction of renal function, the engrafted kidney remaining free of glycolipid deposition for many years [19]. A kidney from a Fabry heterozygote should not be transplanted as it may contain significant storage material resulting in renal dysfunction within a few years [20].

Prophylactic oral anticoagulants are recommended for stroke-prone patients. At an early stage of disease, the use of antiplatelet agents such as aspirin has been suggested as the glycosphingolipid accumulation in vascular endothelium activates platelets and increases angiogenesis. Coronary bypass procedures have been carried out in Fabry patients to treat severe coronary disease. In a 53-year-old female with end-stage cardiomyopathy a heart transplantation was performed. One year after the grafting there were no clinical or histological signs of heart involvement [21].

In summary, Fabry disease is a multisystemic disorder with involvement of several organs including the kidney, heart and cerebrovascular system. Appropriate treatment can be provided only by close cooperation of several specialists, cardiologists, nephrologists and neurologists.

Since until recently in Fabry disease no effective treatment existed which would be able to stop or even reverse the disease process, many attempts have been undertaken to replace the defective α-galactosidase activity with normal enzyme. In a clinical trial, enzyme obtained from human placenta and injected into two Fabry patients was shown to be metabolically active [22]. After early clinical trials clearly demonstrated multiple injections of α-galactosidase A leading to a decrease of plasma globotriaosylceramide, efforts were made to set up large-scale production of the enzyme using molecular genetic techniques. These endeavours ended in clinical studies performed by two different groups. In one double-blind placebo-controlled study conducted by the Mount Sinai Study group [23], enzyme was administered that was produced by Chinese hamster ovary cells (agalsidase beta, Fabrazyme®). This enzyme preparation has been demonstrated to be effective in clearing globotriaosylceramide from the vascular endothelium of the kidney after 20 weeks of treatment, as assessed by light microscopy. This clearance was achieved in 69% (20/29) of the treated patients, but in none of the placebo patients (p < 0.001). This finding was further supported by a significant decrease in globotriaosylceramide inclusions in kidney, heart and skin.

Schiffmann et al. [24] reported the results of a study performed at the National Institutes of Health (NIH). In this randomized trial, an enzyme preparation of human source (agalsidase alfa, Replagal®) was used. The primary endpoint (effect on neuropathic pain) was clearly attained as the treated group showed a consistent and progressive decline in pain scores compared with placebo (p = 0.02). Furthermore, there were documented improvements in secondary endpoints such as creatinine clearance, cardiac conduction and left ventricular mass. Based on the results of these clinical trials, both drugs received approval from the European Union in August 2001.

A positive effect of the enzyme preparation Agalsidase alpha on the heart manifestation was observed by Kampmann and colleagues. They reported a rapid decline of left ventricular mass in 11 female patients who were treated with the enzyme preparation for 24 weeks [25].

In Fabry disease, deposition of the storage material not only occurs in the skin, kidney and heart, but also in the cerebral vasculature, with more localized involvement of central neurons together with dorsal root and autonomic ganglia in the peripheral nervous system. The extent of the cerebrovascular manifestation, best observed in T2-weighted MR sequences, increases with age. Using different methods (PET with [^{15}O] H$_2$O or transcranial Doppler), it has been demonstrated that in patients with Fabry disease, resting regional blood flow is increased and not decreased as one would expect [7]. Cerebral circulation improved significantly on enzyme replacement therapy [8]. The decrease in blood flow velocities may signify a reduced risk of stroke in Fabry disease; this hypothesis can be tested only after long-term treatment.

In patients, the enzyme preparation is generally well tolerated. In the controlled clinical trial, 8 of the 14 patients receiving the enzyme preparation had mild infusion reactions such as chills, facial flushing, nausea or chest pain [24]. These reactions, which appeared about 45 minutes after the infusion, were easily treated with antihistamines and low-dose corticosteroids. Subsequently, the dosage of these drugs could be reduced without any further reactions. Enzyme replacement therapy could be continued.

Fabry disease is a complex disorder showing a wide clinical variation concerning the severity and the onset of symptoms. It has been shown that in individual cases the clinical course cannot be predicted by the level of residual α-galactosidase activity or by analysing the underlying mutation. For these reasons, it is difficult to decide when enzyme replacement therapy should be initiated in asymptomatic patients. The decision becomes even more complicated as there are no surrogate markers which could reliably predict the progress of the disease. Plasma Gb3 levels have been assumed to be a dependable marker, not only indicating the clinical severity but also being a useful tool to follow the effect of enzyme replacement therapy [26]. However, as vascular Gb3 represents only a small proportion of total visceral Gb3 storage, urinary Gb3 levels seem to be a much more valuable marker, particularly regarding kidney function.

Fabry disease has an immense influence on the quality of life, reducing the survival rate in both affected males and females [27]. Dialysis and/or kidney transplantation could not increase life expectancy, underlining the importance of extra-renal manifestations of this lysosomal storage disorder, such as cerebrovascular complications and cardiomyopathy.

Regarding the application of enzyme replacement, many questions still remain:
- What is the indication for enzyme replacement therapy in males and females with Fabry disease: when should treatment be started? When there is heart and kidney dsyfunction or already in childhood to prevent later complications?
- Should every female with Fabry disease be treated?
- Should young children be treated?
- What is the optimal dosage of the enzyme, how often should the drug be given (every week? every month?)

As we gain more insight into Fabry disease and more experience in enzyme replacement therapy, perhaps the next few years will provide answers to these questions.

References

[1] Brady, R. O., A. E. Gal, R. M. Bradley, E. Martensson, A. L. Warshaw, L. Laster: Enzymatic defect in Fabry's disease. Ceramidetrihexosidase deficiency. N. Engl. J. Med. **276**, 1163-1167 (1967).
[2] Garman, S., D. Garboczi: Structural basis of Fabry disease. Mol. Genet. Metab. **77**, 3 (2002).
[3] Garman, S. C., L. Hannick, A. Zhu, D. N. Garboczi: The 1.9 A structure of alpha-N-acetylgalactosaminidase: molecular basis of glycosidase deficiency diseases. Structure **10**, 425-434 (2002).
[4] Luciano, C. A., J. W. Russell, T. K. Banerjee, et al.: Physiological characterization of neuropathy in Fabry's disease. Muscle Nerve **26**, 622-629 (2002).
[5] Friedman, L. S., S. E. Kirkham, J. R. Thistlethwaite, D. Platika, E. H. Kolodny, M. D. Schuffler. Jejunal diverticulosis with perforation as a complication of Fabry's disease. Gastroenterology **86**, 558-563 (1984).
[6] Branton, M. H., R. Schiffmann, S. G. Sabnis, et al.: Natural history of Fabry renal disease: influence of alpha-galactosidase A activity and genetic mutations on clinical course. Medicine **81**, 122-138 (2002).
[7] Moore, D. F., P. Herscovitch, R. Schiffmann: Selective arterial distribution of cerebral hyperperfusion in Fabry disease. J. Neuroimaging **11**, 303-307 (2001).
[8] Moore, D. F., L. T. Scott, M. T. Gladwin, et al.: Regional cerebral hyperperfusion and nitric oxide pathway dysregulation in Fabry disease: reversal by enzyme replacement therapy. Circulation **104**, 1506-1512 (2001).
[9] Tedeschi, G., S. Bonavita, T. K. Banerjee, A. Virta, R. Schiffmann: Diffuse central neuronal involvement in Fabry disease: a proton MRS imaging study. Neurology **52**, 1663-1667 (1999).
[10] Germain, D. P., J. Shabbeer, S. Cotigny, R. J. Desnick: Fabry disease: twenty novel alpha-galactosidase A mutations and genotype-phenotype correlations in classical and variant phenotypes. Mol. Med. **8**, 306-312 (2002).
[11] Whybra, C., C. Kampmann, I. Willers, et al.: Anderson-Fabry disease: clinical manifestations of disease in female heterozygotes. J. Inherit. Metab. Dis. 24, 715-724 (2001).
[12] MacDermot, K. D., A. Holmes, A. H. Miners: Anderson-Fabry disease: clinical manifestations and impact of disease in a cohort of 98 hemizygous males. J. Med. Genet. **38**, 750-760 (2001).
[13] MacDermot, K. D., A. Holmes, A. H. Miners: Anderson-Fabry disease: clinical manifestations and impact of disease in a cohort of 60 obligate carrier females. J. Med. Genet. **38**, 769-775 (2001).
[14] Redonnet-Vernhet, I., J. K. Ploos van Amstel, R. P. Jansen, R. A. Wevers, R. Salvayre, T. Levade: Uneven X inactivation in a female monozygotic twin pair with Fabry disease and discordant expression of a novel mutation in the alpha-galactosidase A gene. J. Med. Genet. **33**, 682-688 (1996).
[15] Poorthuis, B. J., R. A. Wevers, W. J. Kleijer, et al.: The frequency of lysosomal storage diseases in The Netherlands. Hum. Genet. **105**, 151-156 (1999).
[16] Meikle, P. J., J. J. Hopwood, A. E. Clague, W. F. Carey: Prevalence of lysosomal storage disorders. JAMA **281**, 249-254 (1999).
[17] MacDermot, J., K. D. MacDermot: Neuropathic pain in Anderson-Fabry disease: pathology and therapeutic options. Eur. J. Pharmacol. **429**, 121-125 (2001).
[18] Block, F.: Gabapentin zur Schmerztherapie. Nervenarzt **72**, 69-77 (2001).
[19] Mosnier, J. F., C. Degott, J. Bedrossian, et al.: Recurrence of Fabry's disease in a renal allograft eleven years after successful renal transplantation. Transplantation **51**, 759-762 (1991).
[20] Popli, S., Z. V. Molnar, D. J. Leehey et al. : Involvement of renal allograft by Fabry's disease. Am. J. Nephrol. 7, 316-318 (1987).
[21] Cantor, W. J., P. Daly, M. Iwanochko, J. T. Clarke, R. J. Cusimano, J. Butany: Cardiac transplantation for Fabry's disease. Can. J. Cardiol. **14**, 81-84 (1998).

[22] Brady, R. O., J. F. Tallman, W. G. Johnson, et al. : Replacement therapy for inherited enzyme deficiency. Use of purified ceramidetrihexosidase in Fabry's disease. N. Engl. J. Med. **289**, 9-14 (1973).
[23] Eng, C. M., N. Guffon, W. R. Wilcox, et al.: Safety and efficacy of recombinant human alpha-galactosidase A replacement therapy in Fabry's disease. N. Engl. J. Med. **345**, 9-16 (2001).
[24] Schiffmann, R., J. B. Kopp, H. A. Austin, 3rd, et al.: Enzyme replacement therapy in fabry disease: a randomized controlled trial. JAMA **285**, 2743-2749 (2001).
[25] Kampmann, C., M. Ries, F. Baehner, K. S. Kim, M. Bajbouj, M. Beck: Influence of enzyme replacement therapy (ERT) on Anderson Fabry disease associated hypertrophic infiltrative cardiomyopathy (HIC). Eur. J. Pediatr. **161**, R5 (2002).
[26] Eng, C. M., M. Banikazemi, R. E. Gordon, et al.: A phase 1/2 clinical trial of enzyme replacement in Fabry disease: pharmacokinetic, substrate clearance, and safety studies. Am. J. Hum. Genet. **68**, 711-722 (2001).
[27] Gold, K. F., G. M. Pastores, M. F. Botteman, et al.: Quality of life of patients with Fabry disease. Qual. Life Res. **11**, 317-327 (2002).

14 The Anderson-Fabry disease associated cardiomyopathy

C. Kampmann

14.1 Introduction

Anderson-Fabry disease (AFD) (OMIM: 301500) is known to be caused by an X-linked transmitted deficient activity of the lysosomal enzyme α-galactosidase A, leading to an intracellular, lysosomal accumulation of Gb3 (globotriaosylceramide or formerly named ceramidetrihexoside (CTH)), the major substrate of the enzyme [9]. Accumulation occurs in a wide variety of tissues and organs, including the vascular endothelium and the differentiated functional cells [42]. The intracytoplasmatic deposits of Gb3 in the heart are similar to those found in other tissues. Cardiac involvement is frequent due to structural and functional changes of the myocardium, the conduction system, and the valves. Gb3 was found in all cardiac tissues, with the greatest concentrations in the mitral valve and the left ventricular myocardium, while digalactosylceramide, another substrate, was only found to be increased in the lungs and right heart tissues [8]. The Gb3 accumulation leads to an increase in ventricular wall thickness, mitral valve prolapse, and electrocardiogram abnormalities, various degrees of atrioventricular conduction delay, ST-segment, and T-wave abnormalities. Cardiomyopathy rarely is the main manifestation of AFD [11, 12, 40, 47].

Little is known about onset or progression of cardiac involvement in relation to other end-organ manifestations. This may depend on the one hand on the time delay between beginning of heterogenous clinical symptoms and final diagnosis and on the other hand on the rare entity of isolated cardiac manifestation in AFD patients mostly diagnosed post mortem [39]. Onset of first symptoms is at an age of around 9 to 10 years, and the mean time between the first symptoms and diagnosis in classically affected male patients is about 15 years, and in female patients up to 40 years. This depends on the classical clinical signs which are usually first but not always related to AFD in male patients and in female patients because AFD was believed to follow an X-chromosomal recessive transmission. This would indicate that females are normally symptom-free.

The disease manifestations in female heterozygotes, as being obligate carriers, have been reported but were considered to be rare and usually mild; severe and serious manifestations were estimated to be only 1% [9]. A large pedigree examination could show in female carriers multiple and more frequent manifestations of AFD than expected [30, 48]. The

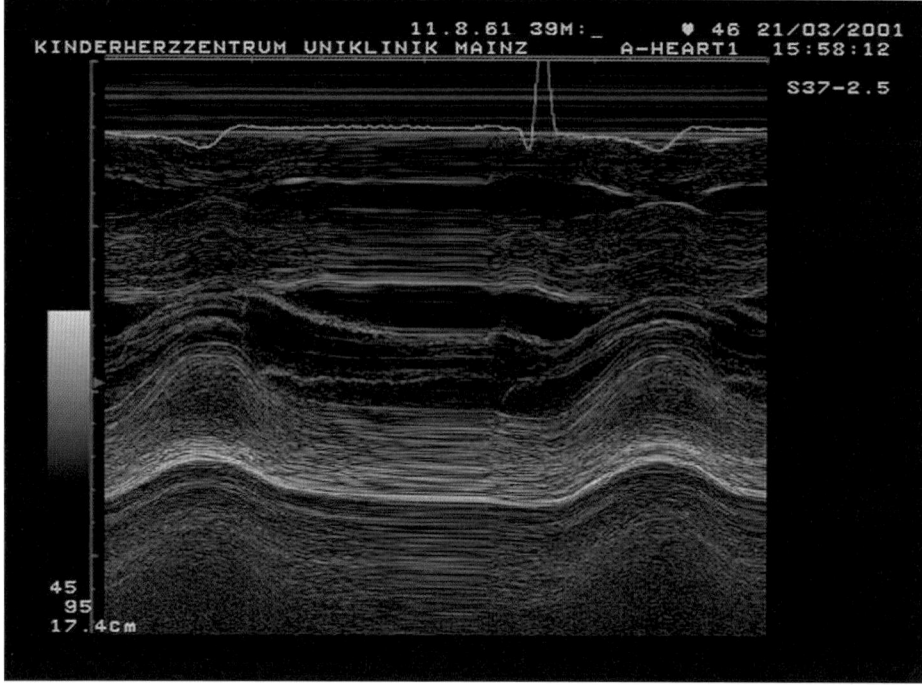

Fig. 1: M-mode echocardiographic tracing of a 39-year-old male patient with severe concentric cardiomyopathy with interventricular and septal thickness greater than 20 mm. Of note, there is also a marked increase in right ventricular anterior wall thickness.

mean age of female affected carriers was around 10 years above that of male hemizygotes [29, 30]. Although there are gender-related differences, cardiac involvement appears to increase with age in both hemizygotes and heterozygotes [25]. Therefore, AFD follows an X-linked rather than a recessive transmission. Thus, careful monitoring for cardiac disease in patients of both sexes is warranted.

14.2 Cardiac manifestation

Early echocardiographic studies on AFD patients revealed an increase in thickness of the posterior left ventricular wall and the interventricular septum, mimicking a hypertrophic cardiomyopathy [2, 18, 45] (Figs. 1, 2), and the severity of the left ventricular hypertrophy correlates well with the severity of the disease [28].

Analogous to other cardiac diseases, hypertrophic changes of the left ventricle are classified according to the distribution of the increased wall thickness and the relation to the end-diastolic diameter of the left ventricle [25]. A normal left ventricle is described as a ventricle with normal mass and a normal ratio of mean wall thickness to end-diastolic diameter. This has to be differentiated from a ventricle with concentric remodelling, in which the left ventricular mass is still within the normal limits, but with increased ratio of mean wall thickness to end-diastolic diameter. The normal calculated left ventricular mass results from a decrease in end-diastolic diameter. Concentric hypertrophy is defined as an increased, supranormal left ventricular mass with increased ratio of mean wall thickness to end-diastolic diameter. Furthermore, an eccentric hypertrophy is defined as increased left ventricular mass with normal ratio of mean wall thickness to end-diastolic diameter, and finally the feature of asymmetrical hypertrophy, which is defined as increased left ventricular mass and a interventricular septal thickness of greater than 15 mm and a ratio of interventricular septal thickness to posterior wall thickness of greater than 1.5. It is important to recognize these different geometrical shapes of the left ventricles, because all of these different shapes are to be seen in Fabry patients and do have their own impact on diastolic function or signs of left ventricular hypertrophy in the ECG [26]. Furthermore, a concentric remodelling is – although the left ventricular mass is still normal – a common finding in AFD, indicating beginning structural changes of the myocardium before a manifest left ventricular hypertrophy has already established. The most common finding in AFD is a concentric hypertrophy (Fig. 3) which counts for almost 50% of the cardiac hypertrophic changes [25, 28]. The second most common finding is a concentric remodelling which is to be seen in about 37% of patients. These patients with concentric remodelling are about

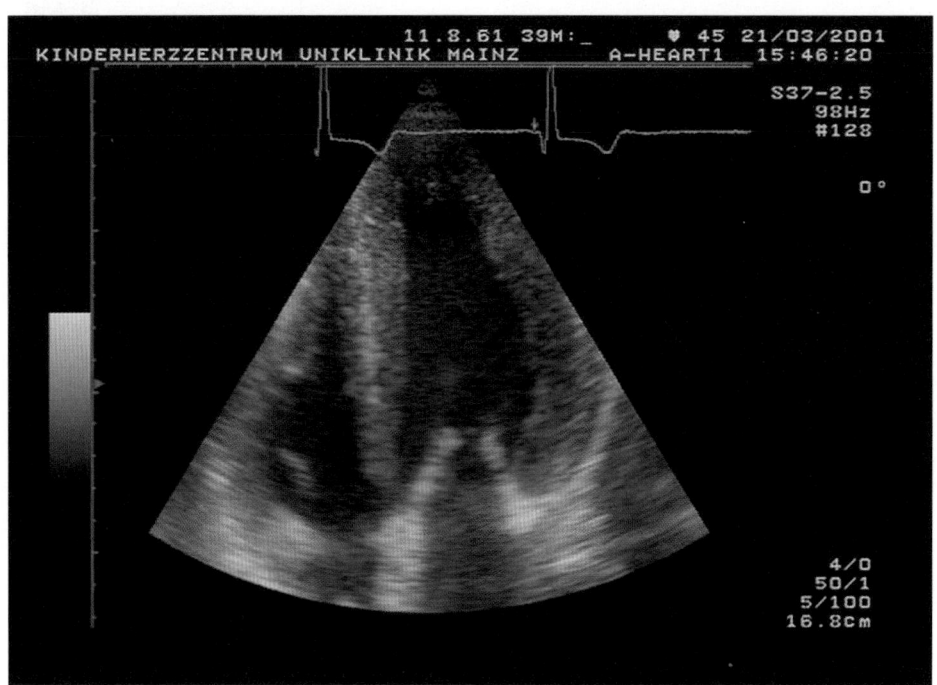

Fig. 2: 2D-echocardiography: 4 chamber view of the left ventricle and the mitral valve in the same 39-year-old male patient. There is a concentric hypertrophy with mitral valve leaflet thickening.

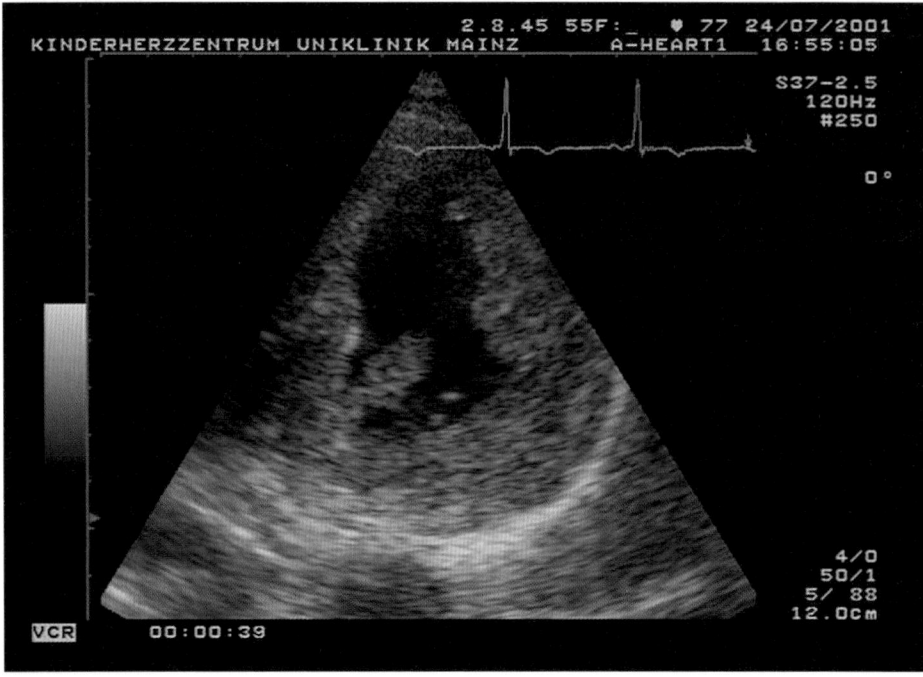

Fig. 3: Modified 2D echocardiographic short axis view of a 55-year-old female patient with severe concentric hypertrophy and thickening of papillary muscle and mitral valve leaflet.

8 to 10 years younger than those with a manifest concentric hypertrophy, leading to the suggestion that concentric remodelling may represent the predominant type of structural abnormality resulting in a concentric hypertrophy in AFD. Eccentric hypertrophy and asymmetrical septal hypertrophy are more seldom but can be seen in about 10 % of patients with cardiac abnormalities [28].

In most patients the end-diastolic volume of the left ventricle decreases with progression of the disease. Diastolic filling is impaired, resulting in a reduction of stroke volume and cardiac output [25]. Additionally, most of the severely affected patients develop brady-arrhythmias or atrial flutter, which contributes to the reduced cardiac output. Thus, severely cardiac affected patients are in a state of compensated low cardiac output.

The mechanisms and pattern of cardiac hypertrophy in AFD are different from those seen in other forms of infiltrative cardiomyopathies (e.g. cardiac amyloidosis). In AFD patients with cardiac involvement there is electrocardiographic evidence of left ventricular hypertrophy marked by increased voltage (Figs. 4, 5), while in patients with other infiltrative cardiomyopathies low voltage is common [7]. Of note, electrocardiographic voltage correlates with left ventricular mass assessed by echocardiography [26]. Increased muscle mass and the absence of interstitial infiltration of the myocardium may be an explanation for non-restrictive filling patterns. The

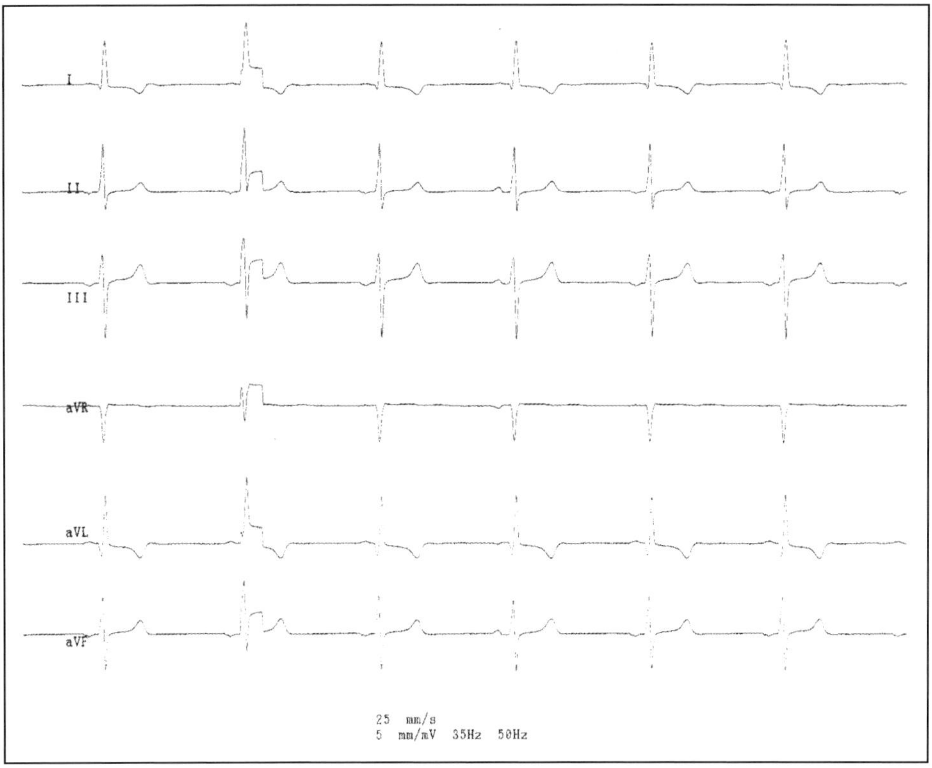

Fig. 4: Typical ECG changes in the same male patient (Figs. 1, 2) with severe cardiomyopathy and brady-arrythmia (heart rate 42 bpm). The patient lost regular sinus-rhythm and changed between sinus-rhythm and ectopic atrial rhythm (25 mm/s; 5 mm/mV).

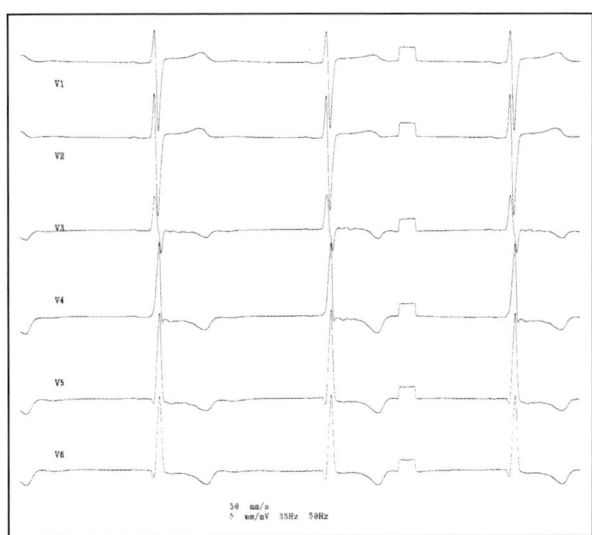

Fig. 5: Typical ECG of the same patient: leads V1 to V6 demonstrate severe biventricular hypertrophy with disturbed repolarisation (50 mm/s; 5mm/mV).

138 Cardiac manifestation

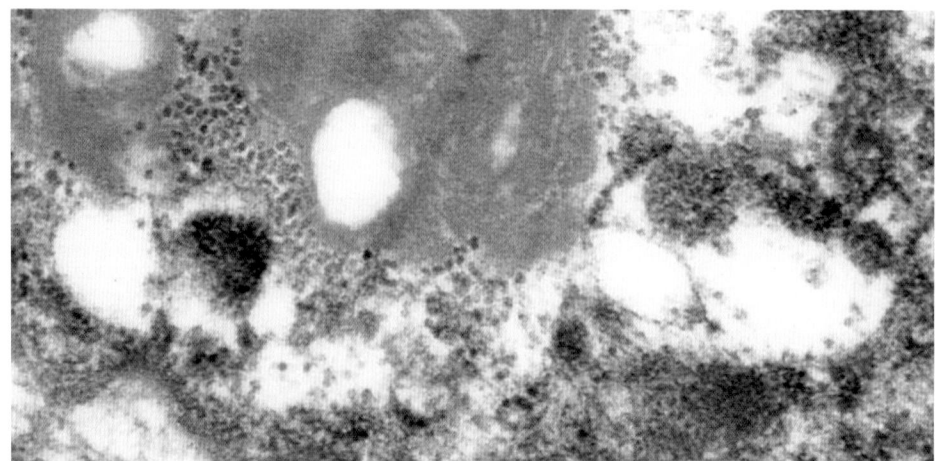

Fig. 6: Electron micrograph of a typical heart specimen of a 45-year-old male patient with severe Anderson-Fabry related cardiomyopathy, showing destruction of myocardial architecture. Of note are the multilammellar myelin bodies of the Gb3 deposits and the large areas of vacuoles.

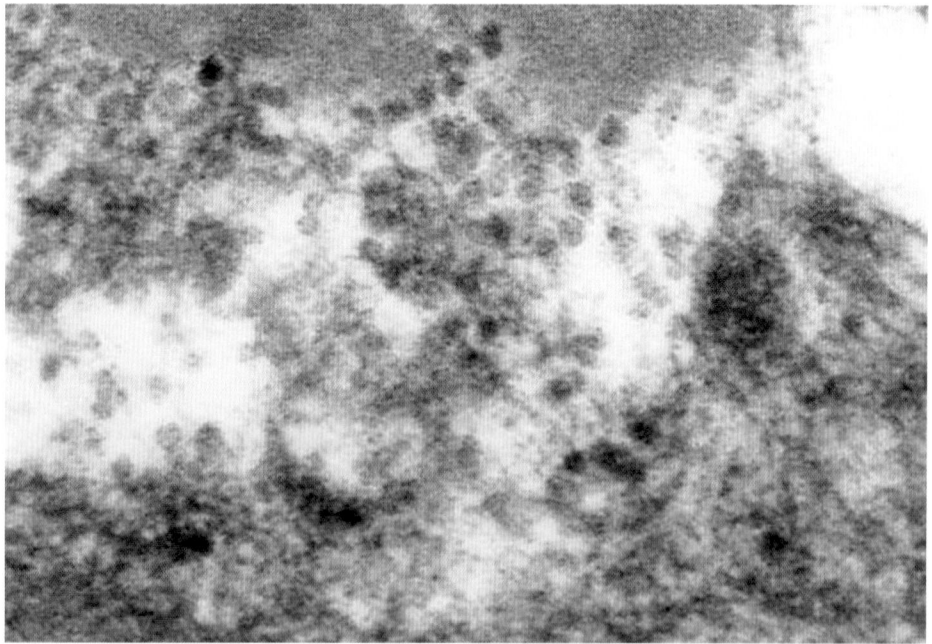

Fig. 7: Electron micrograph of a 45-year-old male patient with valvular aortic stenosis (invasive gradient 38 mmHg) and Anderson-Fabry cardiomyopathy. Here, more vacuolisation than Gb3 accumulation is present. Furthermore, there is a severe myocardial disarray rather than fibrosis.

cardiomyopathy in AFD is a progressive hypertrophic infiltrative cardiomyopathy.

14.2.1 Structural changes of the myocardium

The accumulation of Gb3 can be found in different cardiac tissues, as in the vascular endothelium of the coronary arteries, the valvular fibroblasts, the specific conduction system cells and the myocytes, and therefore affects all cardiac structures, including the musculature of the right ventricle and both atria [8, 15]. It has been shown that the total amount of the Gb3 in the myocardium accounts for around 1% of the myocardial weight (11 mg/g heart wet weight) [11, 47], and cannot be directly responsible for the up to 3 to 4 fold increase in left ventricular mass in end-stage hypertrophy. The progressive lysosomal deposition of the Gb3 leads to a "swelling" of the cardiac cells with an increase in myocardial size. This increase in size results in a change in myocardial cell architecture as comparable to myocardial disarray. Additionally with further progression of the disease there is an increasing number of vacuoles, which can account for more than 60% of the left ventricular mass (Figs. 6, 7) [16]. Whether these vacuoles are the result of auto-phagocytotic progresses with apoptotic destruction of myocardial tissue (in place substitution) or the results of immunological phenomenona superimposed by the presence of Gb3, and therefore in between the myocardial structures, is not clear. It is suggested that there might be an additional neurohumoral influence, confined by increased plasma endothelin-1 levels in AFD patients [27]. Furthermore, sphingoglycolipids and their metabolic products are known to have second messenger functions in a variety of cellular signalling pathways [6]. Lactosylceramide, the catabolic product of Gb3, contributes to atherosclerosis, and was found to stimulate cell proliferation in vitro of aortic smooth muscle cells. Additionally, it mediates the tumour necrosis factor-α (TNF-α) and the intercellular adhesion molecule (ICAM-1) expression. For example, galactosylceramide is important for neuronal cell differentiation and proliferation; lactosylceramide is responsible for cell modulation leading to cell proliferation, binding toxins, activation of NADPH oxidase and so on, while Gb3 is imparted in cell modulation for toxins and apoptosis [6].

14.2.2 Changes in systolic function

Patients with AFD related cardiac hypertrophy have usually normal or increased, age related left ventricular systolic function parameters as measured by a normal or slightly supra-normal endocardial fractional shortening or ejection fraction [2, 22-25, 28]. This is comparable to other hypertrophic myocardial diseases. More advanced measurements of systolic function, as midwall related fractional shortening or the long axis fractional shortening decrease with ongoing hypertrophy. It has been shown that the relation of mean velocity of circumferential fibre shortening to end-systolic wall stress is a more or less load independent parameter of systolic function. Independent of age, patients with AFD related hypertrophy develop with ongoing hypertrophy a decrease in these parameters [25].

Additionally, patients with severe hypertrophy do have an abnormally reduced cardiac output, which is mostly not related to a decreased ejection fraction, but to a decreased stroke volume and abnormally reduced heart rate. The reduced stroke volume is mainly related to the increased ratio of wall thickness to end-diastolic diameter of the left ventricle (the heart is enlarging, but the lumen is becoming smaller) and the reduction in diastolic filling patterns [25].

14.2.3 Changes in diastolic function

To describe the diastolic function is much more difficult than the description of the systolic function, and several different echocardiographic measurements and parameters are available to describe the different filling patterns of the left ventricle. It is of great importance because almost 30 % of congestive heart failure is believed to be related to impaired diastolic function. Diastolic filling patterns are described in 4 major categories with increasing severity: as normal filling, impaired relaxation, pseudo-normalized filling and restrictive filling. Diastolic filling impairs independent of the age of the patient with increasing left ventricular mass, or with increasing left ventricular wall thickness. While it was believed that restrictive left ventricular filling patterns are a common finding in severe Anderson-Fabry related left ventricular hypertrophy, it is a more unusual finding [41, 45]. While in other infiltrative cardiomyopathies like cardiac amyloidosis restrictive filling patterns are usually found in cases with a wall thickness above 15 mm,

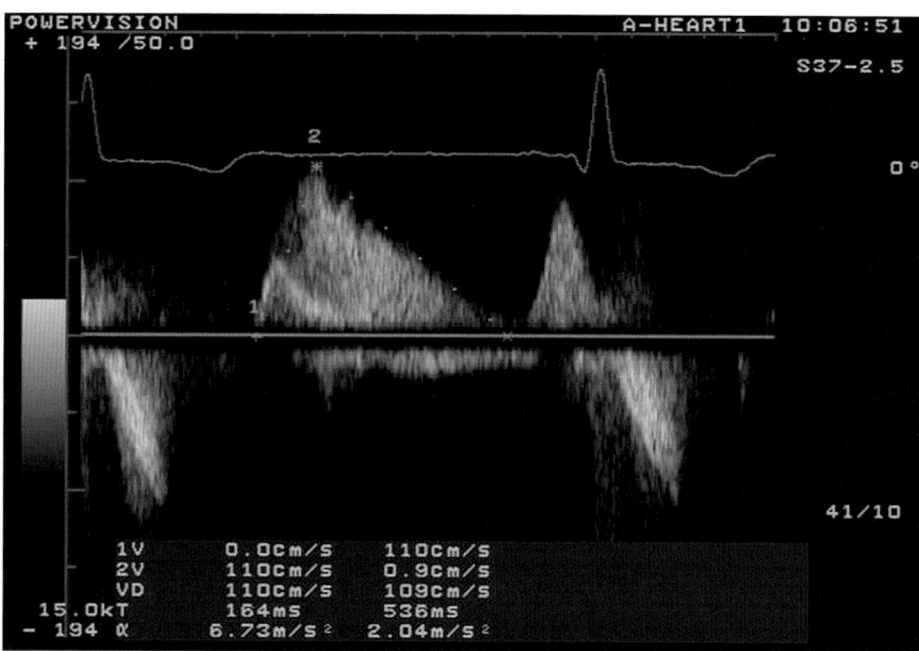

Fig. 8: Doppler tracing across the mitral valve: severe diastolic dysfunction with pseudo-normalized filling patterns in the same 39-year-old male patient (Figs. 1, 2).

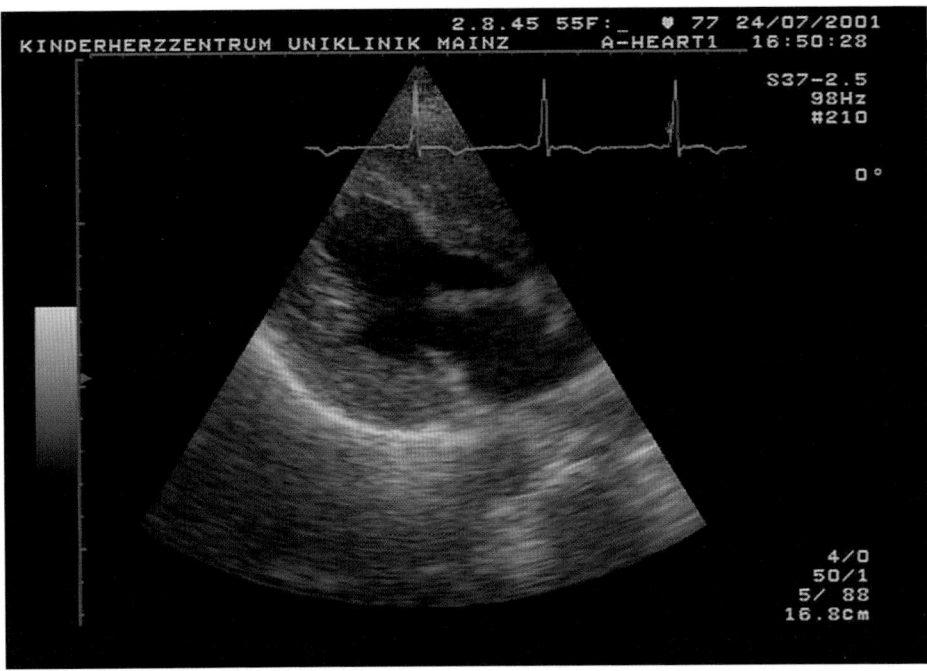

Fig. 9: Modified long axis view of a 55-year-old female patient with severe cardiomyopathy and thickened mitral valve leaflets.

AFD hypertrophy is different. Here, the most common findings in patients with severe cardiomyopathy are an impaired relaxation or pseudo-normalized filling (Fig. 8). Impaired relaxation can even be found in young patients with concentric remodelling. Although restrictive filling is only seldom seen, patients do develop diastolic heart failure even when they show pseudo-normalized filling patterns or impaired relaxation and normal systolic function parameters. The difference between other infiltrative cardiomyopathies and Anderson-Fabry cardiomyopathy in diastolic function properties can be explained by the different location of deposition: in classical infiltrative cardiomyopathies the interstitial deposition of material seems to inhibit the space consuming relaxation and the myofibrils are surrounded by restrictive material. In Anderson-Fabry cardiomyopathy the myofibrils are much less surrounded by restrictive material, but the myocytes themselves are ballooned. With normal myocardial aging the interstitial tissue becomes restrictive because of losing its elastic properties, enhanced by gene expression for fibroblast proliferation. Recently it could be shown that even in patients with normal left ventricular mass, diastolic filling patterns are impaired compared to age matched healthy probands.

Useful parameters to follow the diastolic function in patients with cardiac involvement in Fabry disease are the E to A wave ratio, the E wave deceleration time or the pressure half time (as a function of the deceleration time), the pulmonary venous flow patterns, the systolic anterior movement of the lateral atrioventricular-valve ring and the isovolumic relaxation time. While isovolumic relaxation time is an old fashioned parameter for describing one part of diastolic function, it seems to be one additionally adequate parameter because isovolumic relaxation time increases more or less constantly with ongoing hypertrophy [25]. This again is different from other hypertrophic cardiomyopathies, where isovolumic relaxation time is increased in the mild variants but is in the later stages decreased or pseudo-normalized again.

14.2.4 Valvular involvement

The valvular changes are believed to be caused by the lipid storage and fibrosis of valvular tissue [4], and changes of the leaflet surface do contribute to a higher incidence in endocardial changes. Estimates of valvular involvement in AFD differ significantly. Many authors have suggested that there is a high frequency of mitral valve prolapse [18, 41]; recent reports could not confirm these findings [28]. Nevertheless, minor structural abnormalities of both mitral and aortic valves are frequent. Mitral valve alterations, leaflet thickening (Fig. 9) or prolapse are mostly seen in younger patients, while additional aortic abnormalities appear in the older. Most of the patients with mitral valve abnormalities have thickened papillary muscles due to the myocardial deposition of Gb3, accompanied by mild valvular regurgitation [28]. In the advanced stage with progression of the cardiac involvement and left ventricular hypertrophy, there is a marked aortic root dilatation with changes in the elastic properties of the aortic smooth muscle cells [2, 18, 28]. There are no differences in valvular changes between hemi- and heterozygotes; both genders are affected with more or less the same incidence. Severe alterations of the valves with the need for surgical valve repair are rare. Involvement of the right sided valves as tricuspid and pulmonary valves are extremely rare.

14.2.5 Involvement of the conduction system

Besides accumulation of Gb3 in the ventricular and atrial heart muscle cells and the valves, deposits have been noted throughout the conduction system [1, 15, 20, 31]. This predisposes for rhythm disturbances resulting in tachy- and mostly bradyarrhythmias [38]. Therefore, patients may exhibit an increased susceptibility to supraventricular tachycardia, complete heart blocks, and irregularity of the heart beat (e.g. premature ventricular beats). Therefore, more than 10 % of patients with moderate to severe cardiomyopathy do have pacemaker controlled rhythm in combination with anti-arrhythmic co-medication. In the later stages of AFD, a prolongation of the QRS complex (with a right or left bundle branch block morphology) may also be seen (due to the increase in left ventricular mass) [26, 42]. Atrial fibrillation and diastolic left ventricular dysfunction can rapidly aggravate congestive heart failure.

Furthermore, there are electrocardiographic voltage signs of left ventricular hypertrophy, which correlate well with the echocardiographically assessed left ventricular mass [26]. This differs from other infiltrative cardiomyopathies.

Besides the prolongation of the QRS complex duration time in the absence of bundle branch blocks as a marker of conduction abnormality, there are reports available about a high incidence of a short PR interval (<120 ms) in male patients with severe cardiac hypertrophy [38, 43]. This phenomenon is not constantly seen, and about 15% of patients present with a short PR interval, independent of age.

14.2.6 Coronary artery disease

Both cardiac affected hemi- and heterozygotes complain in more than 50% of anginal chest pain [29, 30]. The electrocardiograms of many cardiac affected patients mimic myocardial infarctions, but without evidence of ischaemic myocardial damage [4, 31].

Endothelial dysfunction may play a dominant role in the development of these symptoms, as endothelial cells from cardiac capillaries are also heavily infiltrated [10]. There might be additional vasospastic neurohumoral stimuli, causing an inadequate vasodilatation of the coronary bed. The ischaemic ECG changes are more likely reflecting left ventricular hypertrophy and the so called "LV strain pattern"; this is presumed to be related to subendocardial ischaemia. Here, subendocardial ischaemia, not infarction, is probably related to oxygen demand outstripping supply, owing to the left ventricular hypertrophy. There is a decreased coronary reserve associated with left ventricular hypertrophy [28]. Whether this suggested chronic reduced oxygen supply enhances myocardial cell proliferation or gene activation is not known. Fixed coronary stenosis and atherosclerosis may be due to hypertension, related to chronic renal failure, and/or associated dyslipidaemia. Furthermore, AFD is a vascular disease and can severely affect the vascular endothelium of the coronary arteries by intracellular deposits of Gb3. There are reports of patients available who have undergone coronary revascularization. In our own patient population of more than 120 males and females with AFD, there was only one 53-year-old female smoker with a severe hypertrophic cardiomyopathy, who had evidence of myocardial infarction and received transluminal angioplasty of the RCA. And of another 3 male smokers older than 45 years with cardiomyopathy and valvular affections, none had angiographic evidence of coronary heart disease although they complained of anginal chest pain. Unfortunately, AFD is usually not diagnosed by cardiologists and even in the case of a patient presenting with a suspected coronary heart disease who receives cardiac catherization and maybe coronary intervention, the diagnosis of AFD can easily be overlooked. Furthermore, since cardiac catherization is not routinely performed in patients with AFD, the true incidence of coronary heart disease compared to age matched patients without AFD remains unclear. Besides that, there is a high incidence of cigarette smokers among AFD patients in Europe, and it is conceivable that nicotine abuse may reduce the neuropathic pain at the onset of symptoms.

Because of the vascular endothelial cell deposition of Gb3, AFD can be described as a vasculopathy with functional end organ affection and damage. Glycosphingolipids in vascular endothelial cells are involved in signalling pathways resulting in atherosclerotic vascular changes [6]. But it is not known why Fabry patients do not suffer from early onset atherosclerosis. The clinical vascular features in severely affected patients are more related to a small vessel disease rather than large vessel damage.

14.3 "Cardiac variant" of AFD

Although AFD is a complex multi-organ disease, there are observations indicating that manifestation can mainly be limited to the heart [11, 36, 40, 44, 47]. It could be shown that the cardiac variant of AFD seems to be more common than previously thought, and around 3 % of male patients with left ventricular hypertrophy suffer from this variant [40]. Preliminary data from an initial screen in a group of unrelated male patients with hypertrophy (either with hypertrophic cardiomyopathy or hypertension) confirmed this finding. In all of these patients the diagnosis of AFD was established by endomyocardial biopsies. Therefore, AFD should be considered as a differential diagnosis of patients with left ventricular hypertrophy, and cardiac biopsy should be performed in each of those suspected patients. Male patients can additionally be identified by measuring the α-galactosidase A activity in serum or plasma. In suspected female patients, while there is usually a normal or slightly reduced enzyme activity, besides cardiac biopsy and the electron micrograph typical changes [46], diagnosis can be established by genetic examination, and prenatal diagnosis is possible. Nevertheless, patients with a so-called cardiac variant have to be examined for other clinical signs, such as angiokeratomas, pain and renal involvement. There are several patients who have been primary diagnosed by ophthalmologists because of their specific eye changes. Furthermore, it has been seen that in the published cases of cardiac variants who were alive, accurate anamnestic and interdisciplinary examination showed that there were other mild organ manifestations and a family history positive for AFD. Therefore, it seems to be more precise to describe the so called cardiac variant as AFD with main manifestation to the heart. Interestingly, those reported patients with the main manifestation to the heart were in general significantly older than those with the classical form of AFD. Furthermore, these patients were only hemizygote male patients who had a reduced α-galactosidase A (around 10% of residual activity) activity comparable to severely affected females [40].

14.4 Gender related differences

AFD is an X-linked disease. It has been believed that AFD follows a recessive transmission mode. Therefore, females were counted as obligate carriers rather than symptomatic patients, and as in other X-linked recessive diseases, females are only very seldom severely affected (1%) [9]. The difference between male and female patients is that the male patients usually have no or severely reduced activity of the α-galactosidase A enzyme, while females usually have normal or slightly reduced activity measured in plasma or leucocytes. Therefore, the final diagnosis of AFD cannot be established by measuring the enzyme activity but just by pedigree and clinical examinations, combined with genetic approaches to detect the mutation on one of the seven exons or introns coding for this defect, or by more invasive procedures, such as endo-myocardial or kidney biopsy. In relation to the heart, females do develop to a much greater extent cardiac involvement, and in fact the most severe cardiomyopathies were seen in female patients [25]. The incidence of general symptoms of AFD in female patients is about 10 years later than in the male population. Of a population of more than 60 females with established diagnosis of AFD about 10 % show a reduced or no enzyme activity as in male patients [30]. Those females had the most severe involvement. Why the other females with almost normal enzyme activity are clinically affected is not totally clear, and further investigation is needed. It can be speculated that according to the Lyon hypothesis the random X-inactivation by each cell is responsible for the large variety of symptoms in the female patients. Indeed, comparing electron micrographs of myocardial tissue from males and females shows in the affected males a accumulation of Gb3 in almost all cells, while in the females affected and non affected cells are present side by side [46]. As usually a substantially large amount of the intracellular produced enzyme is transferred from the intracellular to the extracellular space for re-uptake leading to an intercellular substitution, this mechanism seems not to function sufficiently in females to avoid intracellular deposition of Gb3, although the plasma and leucocyte enzyme activity is normal. This may account for the around 10 year delay of development of symptoms between males and females.

14.5 Onset and progression of cardiomyopathy

Whereas of the classically affected male patients almost all (greater than 90 %) develop a hypertrophic concentric cardiomyopathy at an age of about 28 to 30 years of life, about 60 % of females develop concentric hypertrophy at an age around 38 to 40 years, and the distribution of cardiac involvement is different. Besides concentric hypertrophy, many females show concentric remodelling. Concentric remodelling is a preform of concentric hypertrophy and can cause diastolic dysfunction. Follow-up examinations of patients with cardiac involvement showed that the rate of progression depends on the baseline degree of left ventricular hypertrophy. This means that the progression rate of the left ventricular hypertrophy increases with the degree of baseline hypertrophy. This phenomenon is very interesting, indicating that the progression of the disease follows an exponential function rather than a linear course. Assuming that the accumulation of Gb3 is a constant ongoing process, the exponential development of the hypertrophy is caused by other mechanisms, such as a vicious circle related to a reactive hypertrophy due to ongoing change in the structural myocardial architecture and development of auto-phagocytotic (or apoptotic) vacuoles, resulting in disarray and finally myocardial fibrosis. Additionally, it has been shown that sphingolipids mimic the biological function of cytokines, growth factors and other stress signalling molecules [6], which can interact as second messengers and potentiate the hypertrophic reaction by triggering cell proliferation. Furthermore, this explanation may be used to describe why females develop a cardiomyopathy when the enzyme activity is normal. This reactive increase in myocardial cell size might be the reason why in AFD-related cardiomyopathy the electrocardiogram shows signs of left ventricular hypertrophy. The degree of left ventricular hypertrophy correlates well with classical parameters for left ventricular hypertrophy due to increased afterload, except in those Fabry patients with asymmetrical septal hypertrophy. For the concentric variants of the Fabry cardiomyopathy the cut offs are lower than for hypertensive left ventricular hypertrophy. This seems to be related to the amount of electrically quiet vacuoles.

14.6 Clinical features of cardiac involvement

Classical signs of heart disease are chest pain, angina pectoris, dyspnoea, palpitations, arrhythmias, disabling fatigue and/or syncope. Every one of these symptoms can be related to the leading course of the cardiac problem. Of a registered European population of 336 patients with confirmed diagnosis of AFD, consisting of 154 females aged 40.7 ± 16.4 years and of 182 males aged 35.3 ± 12.6 years, cardiac symptoms reported by the patients themselves were as frequent in heterozygous women as in hemizygous men. At least one of the above-mentioned symptoms was reported by 77 % of male patients and 73 % of female patients. The prevalence of individual symptoms for anginal chest pain was 16% in males and 23 % in females, while dyspnoea was found in 28 % of males and 22 % of females. 16 % of male and 13 % of female patients complained of arrhythmias, and 30 % of male and 39% of female patients complained of disabling fatigue. 9% of males and 4 % of females had at least one episode of syncope. Again, the average age of onset of symptoms was around ten years later in females than in males. Any type of structural involvement in those 336 patients with AFD was 76.6 % in males and 66.7 % in females.

14.7 Cardiac involvement in children with AFD

Although AFD is an inborn error of metabolism, publications about children are rare, although it is known that first clinical symptoms do appear during childhood and adolescence. While in male patients the development of left ventricular hypertrophy is a more constant finding above 28 to 30 years of age, in females the spectrum is more variable and even children or adolescents do have very mild myocardial thickening. In children and adolescents, mitral valve prolapse is more often seen than in an age matched normal population. None of the 19 children and adolescents followed in our institution complain of anginal pain or other clinical features clearly related to cardiac involvement.

14.8 Natural history and death

The natural history of male and female patients is severely influenced by the combination of involvement of the vital organ systems. Besides the involvement of the functional cells of different organ systems, there is a progressive accumulation of Gb3 in the vascular endothelium, leading to ischaemia, especially in the kidney, heart, and brain. It has been shown that the median survival in classically affected male patients was 50 years and in female patients 70 years, which indicates even for the female patients a reduction in life expectancy of around 15 years [29, 30]. The last third of life span is accompanied by severe debilitating manifestations, such as renal insufficiency (more than 30% are on renal replacement therapy – either dialysis or kidney transplantation), hearing loss, neuropathic pain and loss of social integrity. It has been shown for both genders that 24 % of patients die due to cerebrovascular complications. Of 80 reported deaths in pedigree examinations, only 1 patient died due to myocardial infarction. Most of the reported deaths were either related to cerebrovascular complications or sudden death. Taking the wide spectrum of cardiac symptoms into account, the major cause of death of the patients might be related to cardiovascular complications, and mainly to rhythm disturbances or congestive heart failure.

14.9 Treatment

Because of the wide spectrum of clinical manifestations, it is of great importance to cover all different aspects of the patients' symptoms. It has to be shown if long-term treatment with enzyme replacement will resolve all clinical features. Therefore a complex approach is necessary. From a cardiologic point of view there are several different aspects where supportive therapy is mandatory. Bradyarrhythmias, arterial hypertension and more complex rhythm disturbances have to be treated besides enzyme replacement strategies. The control of arterial hypertension is essential to minimize renal, cardiovascular and cerebrovascular disease progression. Prophylaxis with antiplatelet or anticoagulant medication can be important for patients who have had transient ischaemic attacks, strokes or atrial flutter. The most used treatment options are pacemaker implantation, ACE inhibitors and antiarrhythmic drugs such as amiodarone, beta-blockers and digoxin, although it is known that all these preparations may have severe adverse effects in AFD. Amiodarone is known to interact with the lysosomal metabolism, resulting in clinical features comparable to the distinct Fabry signs as with chloroquine or gentamycin, and beta-blockers may reduce heart rate in patients suffering from severe bradyarrhythmias. In the clinical experience of the author, these interventions provide an additional benefit for the patients.

14.9.1 Effects of enzyme replacement therapy on the heart

In recent years novel therapies for AFD have been developed and are now available in Europe [14, 42]. These therapies could have a significant impact on the cardiac manifestations of AFD. A report has been published concerning the improvement of cardiac function in a patient with the cardiac variant of AFD with galactose infusion [16]. In that study, a single patient received three daily infusions of galactose, a competitive inhibitor of the deficient α-galactosidase A, which is believed to enhance the enzyme activity by stabilizing the residual mutant α-galactosidase A enzyme.

Since very large doses of galactose were used in this study, the practicality of this approach is unclear, and the long-term effects of high doses of galactose therapy are unknown. Recently, two different forms of α-galactosidase replacement therapies, one produced in a genetically engineered human cell line – gene activated human – (agalsidase alfa; Replagal©, Transkaryotic Therapies, Cambridge, USA) and the other produced in a Chinese hamster ovary cell line – recombinant – (agalsidase beta; Fabrazyme©, Genzyme, Cambridge, USA), have become available in Europe [3, 14, 17, 42]. Both preparations differ in glycosylation of the protein depending on the originating cell line and the dosage. Human glycosylation (agalsidase alfa) is believed to be an important factor for the intracellular uptake of the enzyme and contributes to a lesser development of IgG antibodies than in non human glycosylation. Besides the type of glycosylation, the human sialic acid (sialylation) seems to be important for the primary

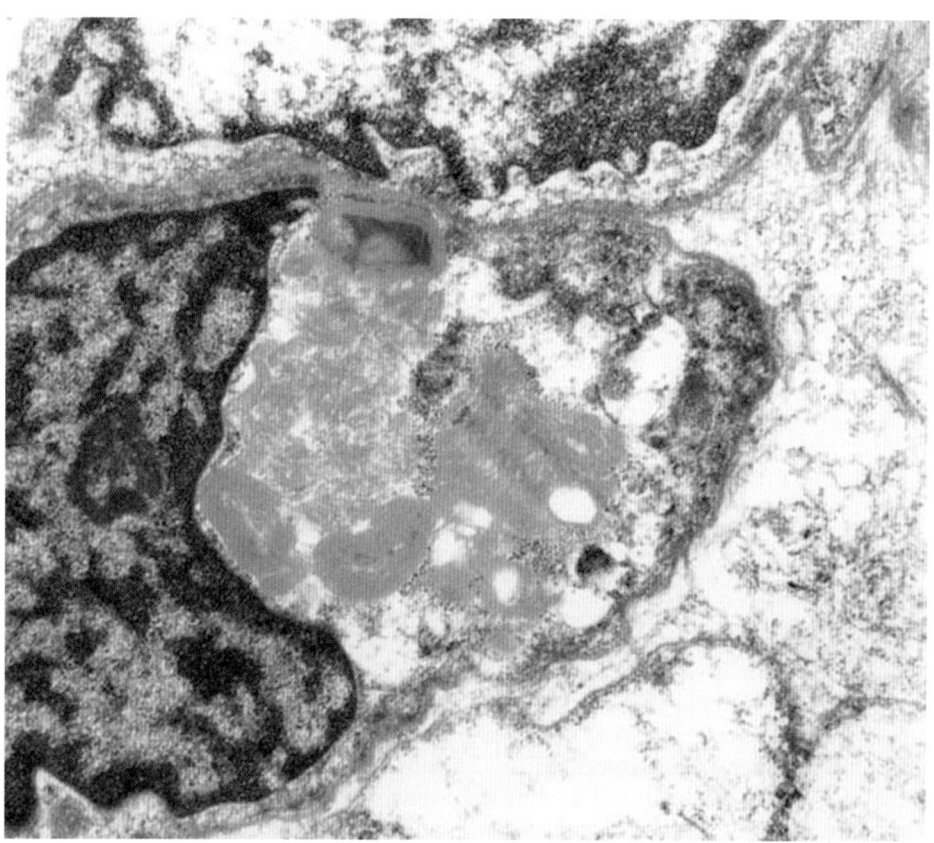

Fig. 10: The same patient as in Fig. 7. Biopsy specimen after 16 weeks (8 infusions) of enzyme replacement therapy with agalsidase alfa. Epithelial cell with intracellular, lysosomal accumulation of globotriaosylceramide (Gb3), but without the typical multilammellar structure, indicating the reduction and resolving of Gb3 as an effect of enzyme replacement therapy.

uptake of the administered enzyme. Both factors may contribute to the difference in the recommended amount of substituted enzyme, although both proteins were found to be structurally and functionally very similar [10]. The recommended dosage of agalsidase beta is five times higher, leading to a five times higher protein load and a six times longer infusion period than agalsidase alfa. While both preparations seem to be in the short-term safe and effective [3, 13, 17, 42], the incidence of anti-anaphylactic co-medication differs; while anti-histamines and ibuprofen or prednisone are recommended before every infusion of agalsidase beta; such medication is only used with agalsidase alfa if patients experience chills or flushes (less than 10% of infusions) [3]. It is reported from the initial studies in male patients for both enzyme replacement therapies that about 80% of the studied patients develop positive IgG antibody titres, which tend to decrease after 6 months of therapy [3]. This IgG antibody development does not lead to discontinuing of enzyme administration or reduced efficacy. This is different from IgE antibody formation. While for agalsidase alfa so far no specific IgE antibody formation has been seen, 5 of 56 patients who received agalsidase beta showed specific IgE titres, which led to a stop of treatment in 4 patients [17].

Two studies describing the safety and efficacy profiles of both enzyme replacement therapies in male patients with AFD have been published [14, 42], and several publications have interpreted these results [3, 5, 17]. Both enzyme preparations have been shown to reverse and reduce the endothelial vascular deposits of Gb3 significantly. In relation to the endomyocardium, the open label phase I/II studies with agalsidase beta showed an increase in Gb3 accumulation of 9.6% [17], while in the double blind placebo controlled phase III study with agalsidase alfa there was a notable (but not significant) decrease in Gb3 deposits in the myocardium by 19 % while in the placebo group the Gb3 increased by 9 % within 6 months [3]. For agalsidase beta, case studies are available reporting a decrease in left ventricular wall thickness and mass.

The study by Schiffmann et al., using agalsidase alfa, showed evidence of an effect on AFD related cardiomyopathy with a significant decrease in QRS duration compared to placebo [42]. Preliminary data of other randomised double blind placebo controlled and open label studies using agalsidase alfa confirmed these findings in reduction in left ventricular mass, and voltage signs, and increase in diastolic function as well as in cardiac output [22]. Furthermore, the decrease and change of intracellular deposits of Gb3 in patients treated with agalsidase alfa was shown by electron microscopy (Fig. 10). For both preparations, much longer studies than half year settings are needed to describe the benefits of enzyme replacement on the heart and kidney. It is conceivable that once the myocardium is structurally ruined by fibrosis or severe disarray, these changes are almost irreversible. In these cases enzyme replacement will hopefully stop the progressive nature of the disease, while in those with just mild affection of the myocardium, substituting the enzyme will possibly result in normalization.

14.9.2 Indication for treatment

The classical form of AFD with multi-organ involvement has an immense influence on quality of life, reducing the median survival rate in both affected males and females by about 15 to 20 years compared to the normal population, and nearly all patients develop renal insufficiency. It is reported that 12% of all Anderson-Fabry patients are today on dialysis in the United States [37]. Dialysis and/or kidney transplantation cannot largely increase life expectancy, underlining the importance of extra renal manifestations such as cardiomyopathy and cerebrovascular complications.

In young patients, the goal of enzyme replacement therapy is to prevent disease, while for the older patients with more advanced and progressed affection, the goal is to both halt disease progression and reverse the underlying pathological abnormalities and the resultant organ dysfunction. Of those patients who are on enzyme replacement with agalsidase alfa it has been shown that the progress of disease can be stopped or slowed down and that the patients improve in pain release, kidney and heart function and correction of brain circulation [5, 22, 33-35, 42].

Enzyme replacement therapies are expensive. Depending on the body weight of the patient, the cost of treatment ranges from € 150 000 to € 200 000 per year per patient. Furthermore, AFD is not as rare as it has been previously suggested, and recent reports assume an estimated incidence of 1 in 40 000 to 1 in 60 000 male live births with classical AFD [10]. Taking into account that females are also affected, the incidence will be much higher. In relation to the population there will be around 2 000 patients with AFD in Germany, of whom about 300 to 500 have already been diagnosed.

However, this cost should be weighed against the expenditures caused by extensive resource utilisation due to dialysis and organ transplantation and loss of productivity and lifespan. The costs of enzyme substitution may be reduced if alternative therapeutic options, such as gene therapy or substrate deprivation, become available.

Enzyme replacement therapy in other lysosomal storage diseases (e.g. Gaucher's disease) have been available for more than 10 years, and it has been shown that preventive enzyme substitution can avoid development of clinical symptoms and reverse symptoms, even of bone metabolism. In classical AFD patients, treatment should be established to avoid or minimize the severe complications of the disease. Furthermore, it is conceivable that in preventive enzyme substitution, the required dosage is lower than that required for reversing of symptoms. This will lower the costs.

References

[1] Bannwart, F.: Fabry's disease. Light and electron microscopic cardiac findings 12 years after successful kidney transplantation. Schweiz. Med. Wochenschr. **112**, 1742-1747 (1982).

[2] Bass, J. L., S. Shrivastava, G. A. Grabowski, R. J. Desnick, J. H. Moller: The M-mode echocardiogram in Fabry's disease. Am. Heart J. **100**, 807-812 (1980).

[3] Beck, M.: Agalsidase alfa – a preparation for enzyme replacement therapy in Anderson-Fabry disease. Expert. Opin. Investig. Drugs **11(6)**, 851-858 (2002).

[4] Becker, A. E., R. Schoorl, A. G. Balk, R. M. van der Heide: Cardiac manifestations of Fabry's disease. Report of a case with mitral insufficiency and electrocardiographic evidence of myocardial infarction. Am. J. Cardiol. **36**, 829-835 (1975).

[5] Brady, R. O., R. Schiffmann: Clinical features of and recent advances in therapy for Fabry disease. JAMA **284**, 2771-2775 (2000).

[6] Chatterjee, S.: Sphingolipids in atherosclerosis and vascular biology. Arterioscler. Thromb. Vasc. Biol. **18**, 1523-1533 (1998).

[7] Cohen, I. S., J. Fluri-Lundeen, T. P. Wharton: Two dimensional echocardiographic similarity of Fabry's disease to cardiac amyloidosis: a function of ultrastructural analogy? J. Clin. Ultrasound **11**, 437-441 (1983).
[8] Desnick, R. J., L. C. Blieden, H. L. Sharp, P. J. Hofschire, J. H. Moller: Cardiac valvular anomalies in Fabry disease. Clinical, morphologic, and biochemical studies. Circulation **54**, 818-825 (1976).
[9] Desnick, R. J., Y. A. Ioannou, C. M. Eng: a-Galactosidase A deficiency: Fabry disease. In: The metabolic and molecular bases of inherited disease. Vol. 3, pp 3733-3774. Scriver, C. H., Beaudet, A. L., Sly, W. S., Valle, D. (eds.). McGraw Hill, New York.
[10] Desnick, R. J., R. Brady, J. Barranger, A. J. Collins, D. P. Germain, M. Goldman, G. Grabowski, S. Packman, W. R. Wilcox: Fabry disease, an under-recognized multisystemic disorder: expert recommendations for diagnosis, management, and enzyme replacement therapy. Ann. Intern. Med. **138**, 338-346 (2003).
[11] Elleder, M., V. Bradova, F. Smid, M. Budesinsky, K. Harzer, B. Kustermann-Kuhn, J. Ledvinova, M. Belohlavek, V. Kral, V. Dorazilova: Cardiocyte storage and hypertrophy as a sole manifestation of Fabry's disease. Report on a case simulating hypertrophic non-obstructive cardiomyopathy. Virchows Arch. A Pathol. Anat. Histopathol. **417**, 449-455 (1990).
[12] Elleder, M., V. Dorazilova, V. Bradova, M. Belohlavek, V. Kral, M. Choura, M. Budesinsky, K. Harzer: Fabry's disease with isolated disease of the cardiac muscle, manifesting as hypertrophic cardiomyopathy. Cas. Lek. Cesk. **129**, 369-372 (1990).
[13] Eng, C. M., M. Banikazemi, R. E. Gordon, M. Goldman, R. Phelps, L. Kim, A. Gass, J. Winston, S. Dikman, J. T. Fallon, S. Brodie, C. B. Stacy, D. Metha, R. Parsons, K. Norton, M. O'Callaghan, R. J. Desnick: A phase 1/2 clinical trial of enzyme relacement in Fabry disease: pharmacokinetic, substrate clearance and safety studies. Am. J. Hum. Genet. **68**, 711-722 (2001).
[14] Eng, C. M., N. Guffon, W. R. Wilcox, D. P. Germain, P. Lee, S. Waldek, L. Caplan, G. E. Linthorst, R. J. Desnick: Safety and efficacy of recombinant human alpha-galactosidase A-replacement therapy in Fabry's disease. N. Engl. J. Med. **345**, 9-16 (2001).
[15] Ferrans, V. J., R. G. Hibbs, C. D. Burda: The heart in Fabry's disease. A histochemical and electron microscopic study. Am. J. Cardiol. **24**, 95-110 (1969).
[16] Frustaci, A., C. Chimenti, R. Ricci, L. Natale, M. A. Russo, M. Pieroni, C. M. Eng, R. J. Desnick: Improvement in cardiac function in the cardiac variant of Fabry's disease with galactose-infusion therapy. N. Engl. J. Med. **345**, 25-32 (2001).
[17] Germain, D. P.: Fabry disease: recent advances in enzyme replacement therapy. Expert. Opin. Investig. Drugs **11(10)**, 1467-1476 (2002).
[18] Goldman, M. E., R. Cantor, M. F. Schwartz, M. Baker, R. J. Desnick: Echocardiographic abnormalities and disease severity in Fabry's disease. J. Am. Coll. Cardiol. **7**, 1157-1161 (1986).
[19] Heltianu, C., G. Costache, K. Azibi, L. Poenaru, M. Simionescu: Endothelial nitric oxide synthase gene polymorphisms in Fabry's disease. Clin. Genet. **61**, 423-429 (2002).
[20] Ikari, Y., K. Kuwako, T. Yamaguchi: Fabry's disease with complete atrioventricular block: histological evidence of involvement of the conduction system. Br. Heart J. **68**, 323-325 (1992).
[21] Ishii, S., S. Nakao, R. Minamikawa-Tachino, R. J. Desnick, J. Q. Fan: Alternative splicing in the a-galactosidase A gene: increased exon inclusion results in the Fabry cardiac phenotype. Am. J. Hum. Genet. **70**, 994-1002 (2002).
[22] Kampmann, C., M. Ries, F. Baehner, K. S. Kim, M. Bajbouj, M. Beck: Influence of enzyme replacement therapy (ERT) on Anderson-Fabry disease associated hypertrophic infiltrative cardiomyopathy (HIC). Eur. J. Pediatr. **161**, R5 (Abstract) (2002).
[23] Kampmann, C., F. Baehner, M. Ries, M. Beck: Cardiac involvement in Anderson-Fabry disease. J. Am. Soc. Nephrol. 13 Suppl. **2**, 147-149 (2002).
[24] Kampmann, C., C. M. Wiethoff, A. Perrot, M. Beck, R. Dietz, K. J. Osterziel: The heart in Anderson-Fabry disease. Z. Kardiol. **214(5)**, 786-795 (2002).
[25] Kampmann, C., F. Baehner, C. Whybra, C. Martin, C. M. Wiethoff, M. Ries, A. Gal, M. Beck: Cardiac manifestation of Anderson-Fabry disease in heterozygous females. J. Am. Coll. Cardiol. **40(9)**, 1668-1674 (2002)
[26] Kampmann, C., C. M. Wiethoff, C. Martin, A. Wenzel, R. Kampmann, C. Whybra, E. Miebach, M. Beck: Electrocardiographic signs of hypertrophy in Fabry disease associated hypertrophic cardiomyopathy. Acta Paediatr. **439**, 21-27 (2002).
[27] Linhart, A., T. Palecek, J. Bultas: Endothelin-1 is associated with advanced clinical symptoms and end-organ involvement in patients with Fabry´s disease. Eur. Heart. J. **21**, 492 (2000).
[28] Linhart, A., T. Palecek, J. Bultas, J. J. Ferguson, J. Hrudova, D. Karetova, J. Zeman, J. Ledvinova, H. Poupetova, M. Elleder, M. Aschermann: New insights in cardiac structural changes in patients with Fabry's disease. Am. Heart J. **139**, 1101-1108 (2000).
[29] MacDermot, K. D., A. Holmes, A. H. Miners: Anderson-Fabry disease: clinical manifestations and impact of disease in a cohort of 98 hemizygote males. J. Med. Genet. **38**, 750-760 (2001).
[30] MacDermot, K. D., A. Holmes, A. H. Miners: Anderson-Fabry disease: clinical manifestations and impact of disease in a cohort of 60 obligate carrier females. J. Med. Genet. **38**, 769-807 (2001).
[31] Mehta, J., N. Tuna, J. H. Moller, R. J. Desnick: Electrocardiographic and vectorcardiographic abnormalities in Fabry's disease. Am. Heart J. **93**, 699-705 (1977).
[32] Miyamura, N., E. Araki, K. Matsuda, R. Yoshimura, N. Furukawa, K. Tsuruzoe, T. Shirotani, H. Kishikawa, K. Yamaguchi, M. Shichiri: A carboxy-terminal truncation of human a-galactosidase A in a heterozygous female with Fabry disease and modification of the enzymatic activity by carboxy-terminal domain. J. Clin. Invest. **98(8)**, 1809-1817 (1996).
[33] Moore, D. F., G. Altarescu, P. Herscovitch, R. Schiffmann: Enzyme replacement reverses abnormal cerebrovascular response in Fabry disease. BMC Neurology **2**, 4 (2002).
[34] Moore, D. F., G. Altarescu, G. S. F. Ling, N. Jeffries, K. P. Frei, T. Weibel, G. Charria-Ortiz, R. Ferri, A. E.

Arai, R. O. Brady, R. Schiffmann: Elevated cerebral blood flow velocities in Fabry disease with reversal after enzyme replacement. Stroke **33**, 525-531 (2002).

[35] Moore, D. F., L. T. C. Scott, M. T. Gladwin, G. Altarescu, C. Kaneski, K. Suzuki, M. Pease-Fye, R. Ferri, R. O. Brady, P. Herscovitch, R. Schiffmann: Regional cerebral hyperperfusion and nitric oxide pathway dysregulation in Fabry disease: reversal by enzyme replacement therapy. Circulation **104**, 1506-1512 (2001).

[36] Ogawa, K., K. Sugamata, N. Funamoto, T. Abe, T. Sato, K. Nagashima, S. Ohkawa: Restricted accumulation of globotriaosylceramide in the hearts of atypical cases of Fabry's disease. Hum. Path. **21**, 1067-1073 (1990).

[37] Pastores, G. M., R. Thadani: Advances in the management of Anderson-Fabry disease: enzyme replacement therapy. Expert. Opin. Biol. Ther. **2(3)**, 325-333 (2002).

[38] Pochis, W. T., J. T. Litzow, B. G. King, D. Kenny: Electrophysiologic findings in Fabry's disease with a short PR interval. Am. J. Cardiol. **74**, 203-204 (1994).

[39] Ries, M., K. Wendrich, C. Whybra, C. Kampmann, A. Gal, M. Beck: Angiokeratoma and pain, but not Fabry's disease: considerations for differential diagnosis. Contrib. Nephrol. **136**, 256-259 (2001).

[40] Sachdev, B., T. Takenaka, H. Teraguchi, C. Tei, P. Lee, W. J. McKenna, P. M. Elliott: Prevalence of Anderson-Fabry disease in male patients with late onset hypertrophic cardiomyopathy. Circulation **105**, 1407-1411 (2002).

[41] Sakuraba, H., Y. Yanagawa, T. Igarashi, Y. Suzuki, K. Suzuki, K. Watanabe, K. Leki, K. Shimoda, T. Yamanaka: Cardiovascular manifestations in Fabry's diesease. Clinical Genetics **29**, 276-283 (1986).

[42] Schiffmann, R., J. B. Kopp, H. A. Austin III, S. Sabnis, D. F. Moore, T. Weibel, J. E. Balow, R. O. Brady: Enzyme replacement therapy in Fabry disease: a randomized controlled trial. JAMA **285**, 2743-2749 (2001).

[43] Senechal, M., D. P. Germain. Fabry disease: a functional and anatomical study of cardiac manifestations in 20 hemizygous male patients. Clin. Genet. **63**, 46-52 (2003).

[44] Sewell, A. C., H. J. Böhles, A. Perrot, K. J. Osterziel: Fabry disease as a potential cause of hypertrophic cardiomyopathy in adult men. J. Inherit. Metab. Dis. **24**, 135 (Abstract) (2001).

[45] Tanaka, H., K. Adachi, Y. Yamashita, H. Toshima, Y. Koga: Four cases of Fabry's disease mimicking hypertrophic cardiomyopathy. J. Cardiol. **18**, 705-718 (1988).

[46] Uchino, M., E. Uyama, H. Kawano, J. Hokamaki, K. Kugiyama, Y. Muratami, M. Yasue, M. Ando: A histochemical and electron microscopic study of skeletal and cardiac muscle from a Fabry disease patient and carrier. Acta Neuropathol. **90**, 334-338 (1995).

[47] von Scheidt, W., C. M. Eng, T. F. Fitzmaurice, E. Erdmann, G. Hübner, E. G. J. Olsen, H. Christomanou, R. Kandolf, D. F. Bishop, R. J. Desnick: An atypical variant of Fabry's disease with manifestations confined to the myocardium. N. Engl. J. Med. **324**, 395-399 (1991).

[48] Whybra, C., C. Kampmann, I. Willers, J. Davis, B. Winchester, J. Kriegsmann, K. Brühl, A. Gal, S. Bunge, M. Beck: Anderson-Fabry disease: Clinical manifestations of disease in female heterozygotes. J. Inherit. Metab. Dis. **24**, 715-724 (2001).

15 Laboratory diagnosis of metabolic diseases presenting with cardiomyopathy

A. C. Sewell

Since the appearance of the 1st edition of this book in 1995, great advances have been made in the diagnosis of inherited metabolic disease, the most important perhaps being the arrival of tandem mass spectrometry. As will be shown later in this chapter, the analysis of acylcarnitines in a dried blood spot will rapidly provide a diagnosis. This has particular relevance for those patients with cardiomyopathy in whom an inherited metabolic disease must be excluded.

Metabolic cardiomyopathies develop within the context of a wide spectrum of pathological conditions [19]. These include an ever-increasing number of inherited metabolic diseases in early childhood affecting the heart and other organs. The paediatrician or paediatric cardiologist presented with a child suffering from cardiomyopathy should be aware of the possibility of an inherited metabolic disease as the underlying cause. This would have far-reaching consequences since it may mean a lethal course and a high risk of recurrence in subsequent siblings. Once a metabolic disorder is diagnosed, appropriate treatment can be initiated and genetic counselling offered.

Since metabolic cardiomyopathies can occur in a wide range of disorders, a suitable diagnostic strategy would be of value in 'screening' those patients in whom an underlying metabolic disorder is suspected. The following groups of inherited metabolic disorders can be delineated:
(1) defects in mitochondrial long-chain fatty acid oxidation,
(2) carnitine deficiency disorders,
(3) respiratory chain defects,
(4) disorders of complex carbohydrate catabolism (glycogenoses, mucopolysaccharidoses and glycoproteinoses),
(5) organic acidaemias and
(6) other diseases such as congenital disorders of glycosylation (CDG).

15.1 Defects in mitochondrial long-chain fatty acid oxidation

The β-oxidation of long-chain fatty acids takes place in the mitochondria, is genetically controlled by nuclear DNA and is linked to the respiratory chain of the cell. Long-chain fatty acids cannot passively pass the mitochondrial membranes and are therefore activated by forming acyl-CoAs. The carnitine transport system (see [8]) provides the means for transport, the long-chain acyl-CoAs then reaching the β-oxidation spiral. Four succesive enzyme reactions then reduce the fatty acid CoA by two carbon atoms liberating electrons and acetyl-CoA.

Since cardiac muscle is reliant on fatty acids as a primary energy source [27], it is logical that any inherited defect in long-chain fatty acid degradation will affect cardiac function. In childhood, the two major inherited defects of long-chain fatty acid oxidation presenting with cardiomyopathy are very long-chain acyl-CoA dehydrogenase (VLCAD) deficiency and long-chain 3-hydroxyacyl-CoA dehydrogenase (LCHAD) deficiency. VLCAD deficiency, a severe form with high mortality in the neonatal period, presents with dilated and hypertrophic cardiomyopathy [1]. LCHAD is part of the mitochondrial tri-functional protein complex [54]. Patients with LCHAD deficiency develop cardiomegaly with a poor contracting left ventricle [52] or hypertrophic cardiomyopathy [13].

Since these metabolic defects are hallmarked by hypoketotic hypoglycaemia, biochemical diagnosis is based upon this important fact [46]. As carnitine is involved in the transport of acyl-CoAs, a secondary carnitine deficiency will ensue. The presence of a dicarboxylic aciduria, when not of dietary origin, can provide further evidence of a β-oxidation defect [18]. Diagnosis of long-chain fatty acid oxidation defects can be achieved by the following:
1. Acylcarnitine analysis in plasma or dried blood spot by tandem mass spectrometry [54]
2. Determination of plasma carnitine, in particular long-chain acylcarnitine, 3-hydroxybutyrate and free fatty acids during a hypoglycaemic episode (blood glucose <40 mg/dl; 2.2 mmol/L).
3. Urinary organic acid analysis demonstrating an unsaturated dicarboxylic aciduria (VLCAD) or a 3-hydroxydicarboxylic aciduria (LCHAD)
4. Analysis of free fatty acids in plasma by gas chromatography [12]

Confirmation can then be accomplished by enzyme analysis in cultured skin fibroblasts followed by mutation analysis [22, 36]. Morphological/histochemical investigations tend to show relatively unspecific features such as lipid droplet accumulation and/or alterations in mitochondrial structure [51].

15.2 Defects in carnitine metabolism

These disorders have been dealt with in a preceding chapter [8]. Since defects in carnitine metabolism can produce cardiomyopathy [57], carnitine analysis in body fluids plays a vital role in any biochemical investigation of patients presenting with such a symptom. Carnitine deficiency may be primary or secondary in nature [48].

The biochemical delineation of carnitine deficiency can be performed as follows:
1. Determination of acylcarnitines in plasma or dried blood spots by tandem mass spectrometry
2. Carnitine analysis in serum/plasma, urine and muscle

Confirmation of carnitine transport defects (systemic carnitine deficiency) can be achieved by carnitine uptake studies in cultured skin fibroblasts [49] followed by mutation analysis. Morphological investigations with the demonstration of lipid myopathy suggest a myopathic form. Secondary disturbances due to CPT 1 or CPT 2 deficiencies can be confirmed by enzyme assay in cultured skin fibroblasts followed by mutation analysis [47].

15.3 Disorders of the respiratory chain

The electron transport chain and oxidative phosphorylation are responsible for mitochondrial energy production. Cardiomyopathy as a prominent clinical feature may be caused by dysfunction or deficiency of respiratory chain complexes or by defects in mitochondrial DNA [55]. The general hallmark of lactic acidosis with increasing blood lactate levels on oral glucose loading may be the first biochemical indication, but cases with no lactate increase are known [43]. Helpful biochemical analyses include:
1. Blood gas and pH
2. Blood lactate before and after glucose load
3. Pyruvate
4. Plasma aminoacids (alanine increase correlating with increased lactate)
5. Urinary lactate/creatinine ratio [15]
6. Urinary aminoacids, phosphate and glucose (Fanconi syndrome in complex IV deficiency [14])
7. Urinary organic acids, in particular the presence of 3-methylglutaconic acid [5, 6]

Confirmation by determination of respiratory chain complex activites in a snap-frozen muscle biopsy is usually mandatory. Histological evidence of ragged red fibres or morphological studies demonstrating increased numbers of enlarged mitochondria are fairly specific [44]. Furthermore, a search for mutations in mitochondrial DNA may help to confirm suspected diagnoses. The term mitochondrial cardiomyopathy has been used to describe patients with either dilated or hypertrophic cardiomyopathy phenotype who have both mitochondrial bioenergetics and mitochondrial DNA defects [38]. Only those patients fulfilling the rigorous criteria of mitochondrial cardiomyopathy can be definitively diagnosed [28].

15.4 Disorders of complex carbohydrate metabolism

Of the known glycogen storage diseases, cardiac manifestations present predominantly in patients with type II (Pompe disease) [53] and type III (debranching enzyme deficiency) [33]. In infantile Pompe disease, the clinical symptoms manifest at birth or shortly thereafter [21] (Figs. 1, 2). Biochemical diagnosis of Pompe disease relies on the determination

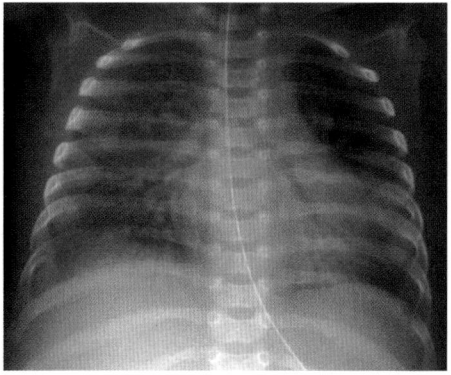

Fig. 1: Chest X-ray film of a patient with Pompe disease (glycogenosis type II) showing massive cardiomegaly.

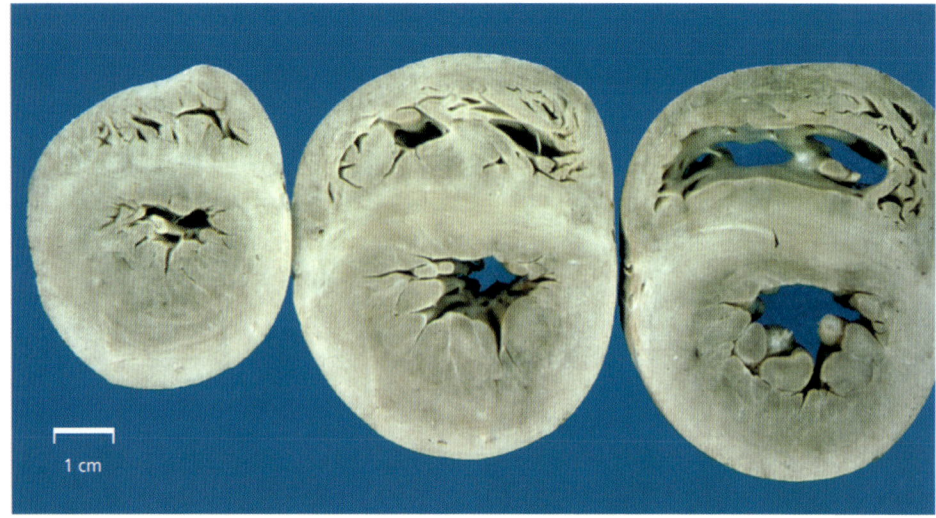

Fig. 2: **Prepared heart of a patient with Pompe disease revealing intense hypertrophic cardiomyopathy.**

of reduced alpha-1,4-glucosidase (acid maltase) activity in lymphocytes, leucocytes and cultured skin fibroblasts [42]. Diagnostic hints may be obtained by examination of peripheral blood smears for the presence of vacuolated lymphocytes and by examination of urine for abnormal oligosaccharides [40]. Histological demonstration of increased glycogen storage provides further evidence. Diagnosis of type III glycogenosis is based solely on enzyme analysis and enzyme immunoreactivity [11]. Apart from one case of phosphorylase-b-kinase deficiency [32], other forms of glycogenosis associated with cardiomyopathy are rare.

Cardiac problems are well known in some forms of mucopolysaccharidosis (types I Hurler, II Hunter and VI Maroteaux-Lamy) but mainly concern valve abnormalities due to tissue deposition of excess glycosaminoglycans [34]. Of the glycoproteinoses, patients with GM1 gangliosidosis, GM2 gangliosidosis type Sandhoff and sialidosis are known to have cardiomyopathy [7, 37]. Single case reports have demonstrated cardiac involvment in I-cell disease [35], Niemann-Pick type A [56] and galactosialidosis [41, 45]. Among the sphingolipidoses, cardiomyopathy is known to occur in Gaucher disease [16, 26] and Fabry disease [23].

Biochemical diagnosis of mucopolysaccharidoses and glycoproteinoses is dependent upon clinical suspicion and the radiological demonstration of dysostosis multiplex. Abnormal urinary glycosaminoglycans [39] and oligosaccharides [40] are pathognomic. Increased numbers of vacuolated lymphocytes in peripheral blood provide further evidence and enzyme determinations in blood, tissues and cultured skin fibroblasts confirm the diagnosis. Clinical suspicion of Gaucher disease and Fabry disease can de confirmed by direct enzyme assay in appropriate samples (blood or tissue) and may be complemented by mutation analysis [10].

15.5 Organoacidopathies

Cardiological symptoms are known in a number of organic acid disorders. In a retrospective study of patients with propionic aciduria, 6 of 19 patients developed a cardiomyopathy [30]. Development in 2 patients was extremely rapid, but metabolic decompensation was not considered to be the cause. Secondary carnitine insufficiency was also considered unlikely as 3 out of 4 patients died despite carnitine therapy and 2 patients improved without carnitine supplemenatation [30]. A subgroup of patients with an X-linked form of 3-methylglutaconic aciduria all had dilated cardiomyopathy and growth retardation with a severe and lethal progression [25]. Other single case reports have described cardiological problems in methylmalonic aciduria [50], malonic aciduria [31, 58], β-ketothiolase deficiency [20], D-2-hydroxyglutaric aciduria [4] and hydroxymethylglutaryl-CoA lyase deficiency [17].

Biochemical diagnosis relies on the demonstration of increased pathological organic acids in urine and plasma. Secondary carnitine insufficiency can be confirmed by acylcarnitine analysis in plasma or dried blood spots by tandem mass spectrometry [9]. Confirmation of the disease should be accomplished by enzyme assay in suitably available tissues.

15.6 Others

Of the other diseases presenting with cardiomyopathy, congenital disorders of glycosylation (CDG) are of importance and are definitely under-diagnosed [29]. Diagnosis is accomplished by demonstrating an abnormal sialylated transferrin pattern in serum using isoelectric focussing together with low levels of other serum glycoproteins such as AT III or transcortin. Patients with phosphoenolpyruvate carboxykinase deficiency have been described with cardiac symptoms [3]. Diagnostic hints are fasting hypoglycaemia without reaction to glucagon and lactic acidaemia, confirmation is by enzyme assay in liver tissue or cultured skin fibroblasts.

In tyrosinaemia type 1, cardiomyopathy is an important clinical manifestation. A recent study demonstrated that of 17 children with tyrosinaemia type 1, 40% had hypertrophic cardiomyopathy which normalised irrespective of the mode of treatment [2]. Diagnosis of tyrosinaemia type 1 is readily obtained by a combination of plasma aminoacid analysis and urinary organic acid analysis.

The aim of this article was to describe laboratory diagnostic aspects of metabolically related cardiac manifestation and to provide a framework within which the most important diseases could be identified. The strategy illustrated (Fig. 3) may go some way to provide such a framework. The determination of lactate, acylcarnitines, carnitine, aminoacids and organic acids will provide diagnostic information for a variety of disorders. Special analyses are required for glycogenoses, CDG, Fabry disease and Gaucher disease. Most lysosomal storage disorders rely on concrete

158 Others

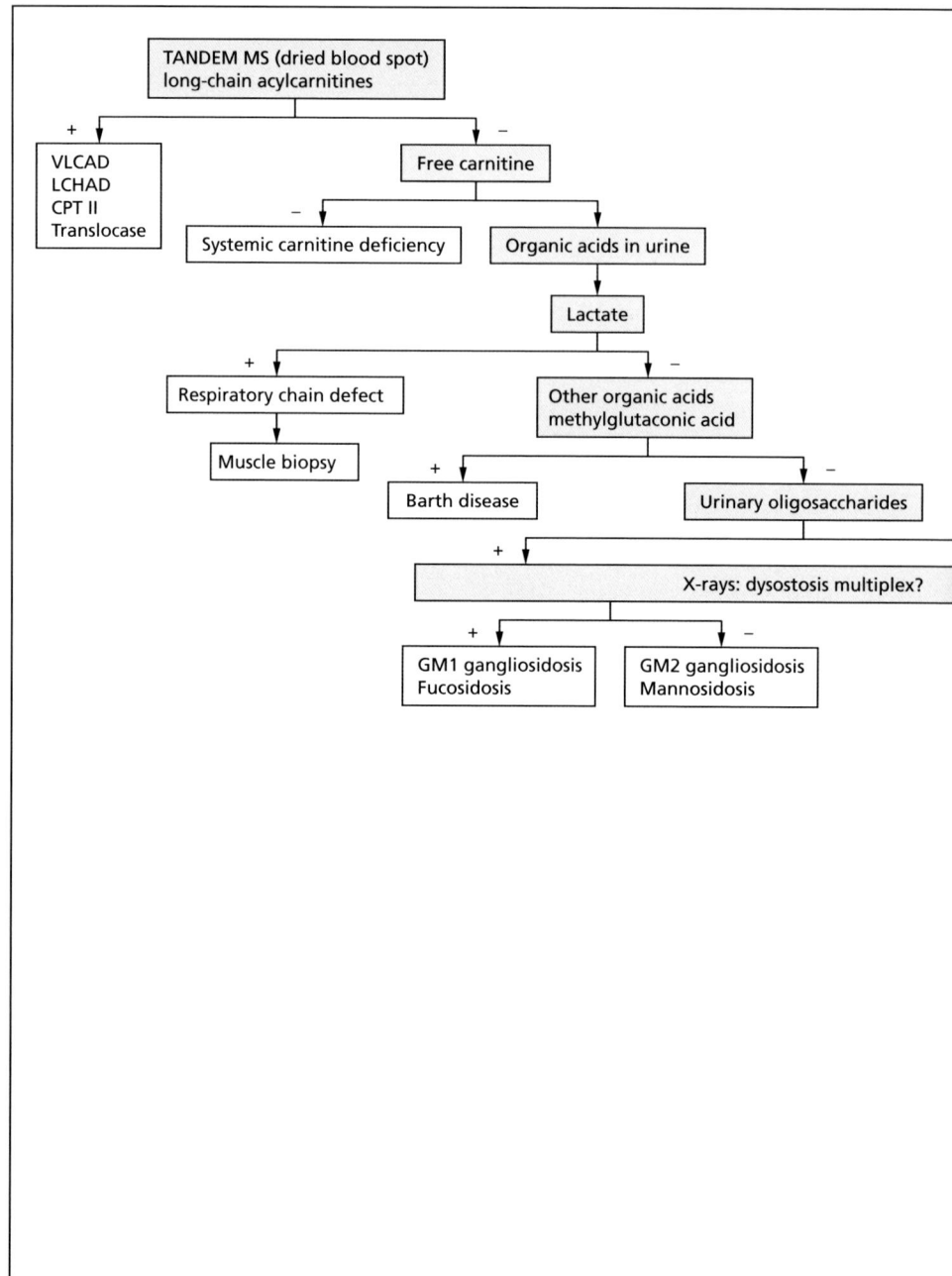

Fig. 3: A flowsheet of laboratory screening for metabolic cardiomyopathy.

Laboratory diagnosis of metabolic diseases 159

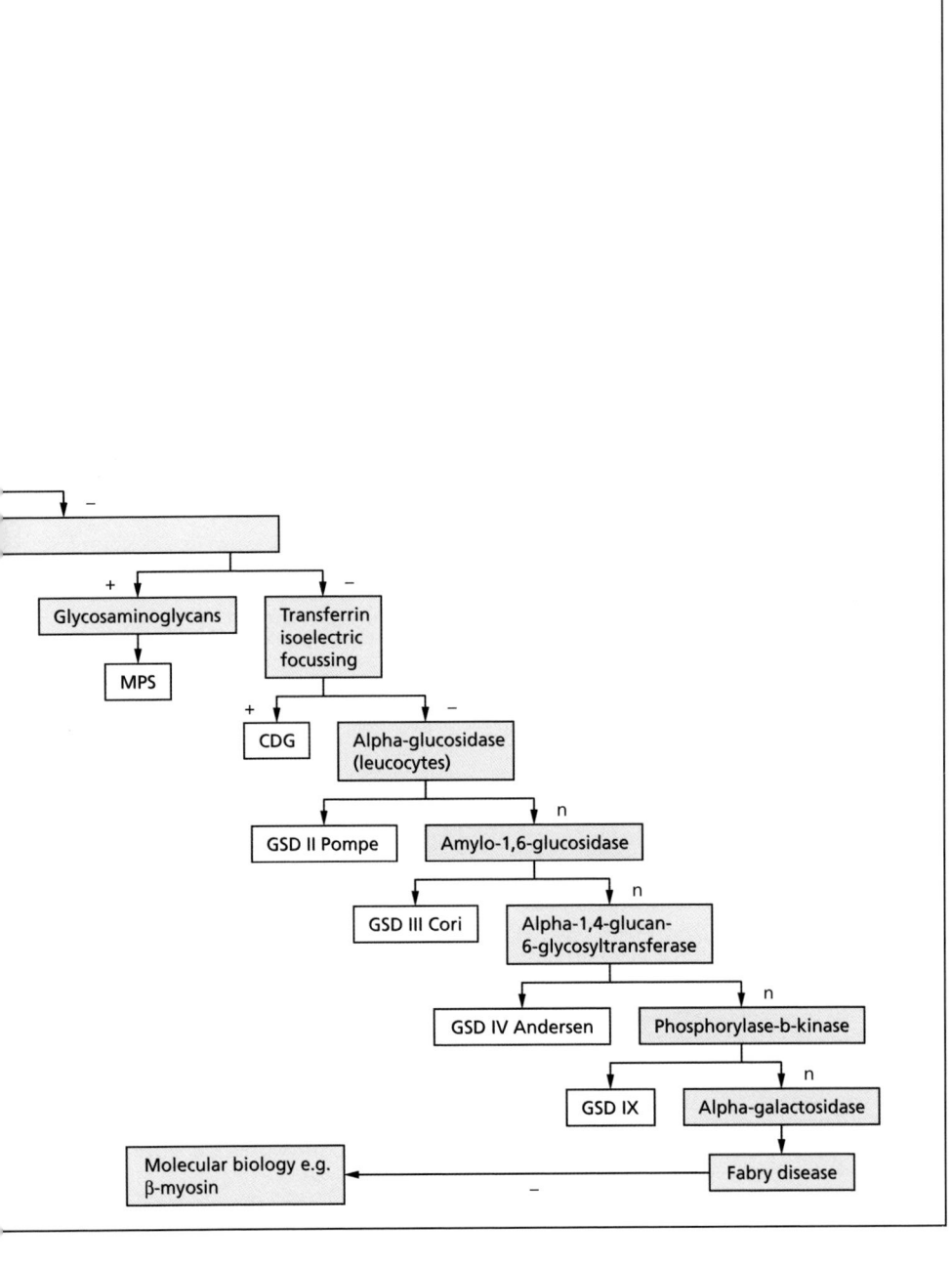

clinical symptoms together with radiological evidence - only then is screening of urine for excess glycosaminoglycan and oligosacharide excretion warranted. Evidence is now emerging that inherited disorders are a major cause of cardiomyopathy [19, 24]. Since this symptom leads to morbidity and mortality in both children and adults, the delineation of cardiomyopathy and its molecular basis is an important initial step in developing improved methods of screening and therapy for patients with this disease.

References

[1] Andresen, B. S., S. Olpin, B. J. M. Poorthuis et al.: Clear correlation of genotype with disease phenotype in very-long-chain acyl-CoA dehydrogenase deficiency. Am. J. Hum. Genet. **64**, 479-494 (1999).
[2] Arora, N., P. J. McKiernan, O. Stumper, J. Wright: Cardiomyopathy in tyrosinaemia type 1 is associated with a good long-term outcome. J. Inherit. Metab. Dis. **24** (Suppl. 1), 33 (2001).
[3] Baerlocher, K.: Disorders of gluconeogenesis. In: Inborn metabolic diseases. Diagnosis and treatment, pp. 113-123. Fernandes, J., J.-M. Saudubray, K. Tada (eds.). Springer, Berlin (1990).
[4] Baker, N. S., H. B. Sarnat, R. M. Jack et al.: D-2-hydroxyglutaric aciduria: hypotonia, cortical blindness, seizures, cardiomyopathy, and cylindrical spirals in skeletal muscle. J. Child. Neurol. **12**, 31-36 (1997).
[5] Bennett, M. J., W. G. Sherwood, K. M. Gibson, A. B. Burlina: Secondary inhibition of multiple NAD-requiring dehydrogenases in respiratory chain complex I deficiency: possible metabolic markers for the primary defect. J. Inherit. Metab. Dis. **16**, 560-562 (1993).
[6] Besley, G. T. N., M. Lendon, D. M. Broadhead et al.: Mitochondrial complex deficiencies in a male with cardiomyopathy and 3-methylglutaconic aciduria. J. Inherit. Metab. Dis. **18**, 221-223 (1995).
[7] Blieden, L. C., R. J. Desnick, J. B. Carter et al.: Cardiac involvement in Sandhoff's disease. Am. J. Cardiol. **34**, 83-88 (1974).
[8] Böhles, H. J. (this volume)
[9] Chace, D. H., T. A. Kalas, E. W. Naylor: The application of tandem mass spectrometry to neonatal screening for inherited disorders of intermediary metabolism. Annu. Rev. Genomics Hum. Genet. **3**, 17-45 (2002).
[10] Charrow, J., H. C. Andersson, P. Kaplan et al.: The Gaucher registry, demographics and disease characteristics of 1698 patients with Gaucher disease. Arch. Intern. Med. **160**, 2835-2843 (2000).
[11] Coleman, R. A., H. S. Winter, B. Wolf et al.: Glycogen storage disease type III (glycogen debranching enzyme deficiency): correlation of biochemical defects with myopathy and cardiomyopathy. Ann. Intern. Med. **116**, 896-900 (1992).
[12] Costa, C. G., L. Dorland, U. Holwerda et al.: Simultaneous analysis of plasma free fatty acids and their 3-hydroxy analogs in fatty acid β-oxidation disorders. Clin. Chem. **44**, 463-471 (1998).
[13] Das, A. M., R. Fingerhut, R. J. A. Wanders, K. Ullrich: Secondary respiratory chain defect in a boy with long-chain 3-hydroxyacyl-CoA dehydrogenase deficiency: possible diagnostic pitfalls. Eur. J. Pediatr. **159**, 243-246 (2000).
[14] DiMauro, S., E. Bonilla, M. Zeviani et al.: Mitochondrial myopathies. Ann. Neurol. **17**, 521-538 (1985).
[15] Dunger, D. B., J. V. Leonard: An evaluation of urine lactate for detection of inborn errors of metabolism. J. Inherit. Metab. Dis. **7**, 111-113 (1984).
[16] Edwards, W. D., H. P. Hurdey, J. R. Partin: Cardiac involvement in Gaucher's disease documented by right ventricular endomyocardial biopsy. Am. J. Cardiol. **52**, 654 (1983).
[17] Gibson, K. M., C. F. Lee, S. B. Cassidy et al.: Fatal cardiomyopathy associated with 3-hydroxy-3-methylglutaryl-coenzyme A (HMG-CoA) lyase deficiency. Abstract 31st SSIEM Symposium, Manchester W21 (1993).
[18] Gregersen, N., S. Kolvraa, K. Rasmussen et al.: General (medium-chain) acyl-CoA dehydrogenase deficiency (non ketotic dicarboxylic aciduria): quantitative urinary excretion pattern of 23 biologically significant organic acids in 3 cases. Clin. Chim. Acta **132**, 181-191 (1983).
[19] Guertl, B., C. Noehammer, G. Hoefler: Metabolic cardiomyopathies. Int. J. Exp. Pathol. **81**, 349-372 (2000).
[20] Henry, C. G., A. W. Strauss, J. P. Keating, H. E. Hillman: Congestive cardiomyopathy associated with beta-ketothiolase deficiency. J. Pediatr. **99**, 754-757 (1981).
[21] Hers, H. G., F. Van Hoof, T. de Barsy: Glycogen storage diseases. In: The metabolic basis of inherited disease, vol 1, pp. 425-452. Scriver, C. R., A. L. Beaudet, W. S. Sly, D. Valle (eds.). McGraw-Hill, New York (1989).
[22] Ijlst, L., S. Usikubo, T. Kamijo et al.: Long-chain 3-hydroxyacyl-CoA dehydrogenase deficiency: high frequency of the G1528C mutation with no apparent correlation with the phenotype. J. Inherit. Metab. Dis. **18**, 241-244 (1995).
[23] Kampmann, C. (this volume)
[24] Kelley, D. P., A. W. Strauss: Inherited cardiomyopathies. N. Engl. J. Med. **330**, 913-919 (1994).
[25] Kelley, R. I., J. P. Cheatham, B. J. Clark et al.: X-linked dilated cardiomyopathy with neutropenia, growth retardation and 3-methylglutaconic aciduria. J. Pediatr. **119**, 738-747 (1991).
[26] Laks, Y., J. Passwell: The varied clinical and laboratory manifestations of type II Gaucher disease. Acta Paediatr. Scand. **76**, 378-380 (1987).
[27] Lopaschuk, G. D., D. D. Belke, J. Gamble et al.: Regulation of fatty acid oxidation in the mammalian heart in health and disease. Biochim. Biophys. Acta **213**, 263-276 (1994).
[28] Marin-Garcia, J., R. Ananthakrishnan, M. J. Golden-

thal, M. E. Pierpont: Biochemical and molecular basis for mitochondrial cardiomyopathy in neonates and children. J. Inherit. Metab. Dis. **23**, 625-633 (2000).
[29] Marquardt, T. (this volume)
[30] Massoud, A. F., J. V. Leonard: Cardiomyopathy in propionic acidaemia. Eur. J. Pediatr. **152**, 441-445 (1993).
[31] Matalon, R., K. Michaels, R. Kaul et al.: Malonic aciduria and cardiomyopathy. J. Inherit. Metab. Dis. **16**, 571-573 (1993).
[32] Mizuta, K., E. Hashimotot, A. Tsutou et al.: A new type of glycogen storage disease caused by deficiency of cardiac phosphorylase kinase. Biochem. Biophys. Res. Commun. **119**, 582-587 (1984).
[33] Moses, S. W., K. L. Wandermann, A. Myroz, M. Frydman: Cardiac involvement in glycogen storage disease type III. Eur. J. Pediatr. **148**, 764-766 (1989).
[34] Neufeld, E. F., J. Muenzer: The mucopolysaccharidoses. In: The metabolic and molecular basis of inherited disease, 7th ed., pp. 2465-2494. Scriver, C. R., Beaudet, A. L., Sly, W. S., Valle, D. (eds.). McGraw-Hill, New York (1995).
[35] Okada, S., M. Owada, T. Sakiyama et al.: I-cell disease: clinical studies of 21 Japanese cases. Clin. Genet. **28**, 207-215 (1985).
[36] Pons, R., P. Cavadini, S. Baratta et al.: Clinical and molecular heterogeneity in very-long-chain acyl-coenzyme A dehydrogenase deficiency. Pediatr. Neurol. **22**, 98-105 (2000).
[37] Rosenberg, H., T. C. Frewen, M. D. Li et al.: Cardiac involvement in diseases characterized by β-galactosidase deficiency. J. Pediatr. **106**, 78-80 (1985).
[38] Schon, E. A., E. Bonilla, S. DiMauro: Mitochondrial DNA mutations and pathogenesis. J. Bioenerg. Biomembr. **29**, 131-149 (1997).
[39] Sewell, A. C.: Urinary screening for disorders of heteroglycan metabolism. Results of 10 years experience with a comprehensive system. Klin. Wochenschr. **66**, 48-53 (1988).
[40] Sewell, A. C.: Urinary oligosaccharides. In: Techniques in diagnostic human biochemical genetics. A laboratory manual, pp. 219-231. Hommes, F. A. (ed.). Wiley-Liss, New York (1991).
[41] Sewell, A. C., B. F. Pontz, D. Weitzel, C. Humburg: Clinical heterogeneity in infantile galactosialidosis. Eur. J. Pediatr. **146**, 528-531 (1987).
[42] Shin, Y. S., W. Endres, J. Unterreithmeier et al.: Diagnosis of Pompe's disease using leukocyte preparations. Kinetic and immunological studies of 1,4-alpha-glucosidase in human fetal and adult tissues and cultured cells. Clin. Chim. Acta **148**, 9-19 (1985).
[43] Smeitink, J. A. M., W. G. Blok, J. C. Fischer et al.: Normal oral glucose lactate stimulation test in a patient with cytochrome c oxidase deficiency. Enzyme Protein **47**, 6 (1993).
[44] Stadhouders, A. M., R. C. A. Sengers: Morphological observations in skeletal muscle from patients with a mitochondrial myopathy. J. Inherit. Metab. Dis. **10**, 62-80 (1987).

[45] Strisciuglio, P., W. S. Sly, W. E. Dodson et al.: Combined deficiency of beta-galactosidase and neuraminidase: natural history of the disease in the first 18 years of an American patient with late infantile onset form. Am. J. Med. Genet. **37**, 573-577 (1990).
[46] Taroni, F., G. Uziel: Fatty acid mitochondrial oxidation and hypoglycaemia in children. Curr. Opin. Neurol. **9**, 477-485 (1996).
[47] Taroni, F., E. Verderio, S. Fiorucci et al.: Molecular characterisation of inherited carnitine palmitoyltransferase II deficiency. Proc. Natl. Acad. Sci. USA **89**, 8429-8433 (1992).
[48] Tein, I., S. DiMauro: Primary systemic carnitine deficiency manifested by carnitine-responsive cardiomyopathy. In: L-carnitine and its role in medicine; from function to therapy, pp. 155-184. Ferrari, R., DiMauro, S., Sherwood, G. (eds.). Academic Press, New York (1992)
[49] Tein, I., D. Devivo, F. Bierman et al.: Impaired skin fibroblast uptake in primary systemic carnitine deficiency manifested by childhood carnitine-responsive cardiomyopathy. Pediatr. Res. **28**, 247-255 (1990).
[50] Thompson, G. N., D. M. Danks, D. B. Edis: Cardiomyopathy as a complication of methylmalonic acidaemia. 25th SSIEM Annual Symposium, Birmingham, p. 43 (1990).
[51] Treem, W. R., C. A. Witzleben, D. A. Piccoli et al.: Medium-chain and long-chain acyl-CoA dehydrogenase deficiency: clinical, pathologic and ultrastructural differentiation from Reye's syndrome. Hepatology **6**, 1270-1278 (1986).
[52] Tyni, T., A. Palotie, L. Viikina et al.: Long-chain 3-hydroxyl-coenzyme A dehydrogenase deficiency with the G1528C mutation: clinical presentation of thirteen patients. J. Pediatr. **130**, 67-76 (1997).
[53] Ullrich, K., H. Hahn-Ullrich, H.-G. Koch: Cardiac alterations in glycogenoses, glycoproteinoses, gangliosidoses and carbohydrate deficient glycoprotein syndrome. In : Metabolic cardiomyopathy. 1st ed., pp. 99-114. Böhles, H., Hofstetter, R., Sewell, A. C. (eds.). Wiss. Verlagsgesellschaft, Stuttgart (1995).
[54] Van Hove, J. L. K., S. G. Kahler, M. D. Feezor et al.: Acylcarnitines in plasma and blood pots of patients with long-chain 3-hydroxyacyl-coenzyme A dehydrogenase deficiency. J. Inherit. Metab. Dis. **23**, 571-582 (2000).
[55] Wallace, D. C.: Mitochondrial defects in cardiomyopathy and neuromuscular disease. Am. Heart J. **139**, 70-85 (2000).
[56] Westwood, M.: Endocardial fibroelastosis and Niemann-Pick disease. Br. Heart J. **39**, 1394-1396 (1977).
[57] Winter, S. C., N. R. M. Buist: Cardiomyopathy in childhood, mitochondrial dysfunction and the role of carnitine. Am. Heart J. **139**, 63-69 (2000).
[58] Yano, S., L. Sweetman, D. Thorburn et al.: A new case of malonyl coenzyme coenzyme A decarboxylase deficiency presenting with cardiomyopathy. Eur. J. Pediatr. **156**, 382-383 (1997).

Authors

Prof. Dr. M. Beck
Children's Hospital
University of Mainz
Langenbeckstr. 1
D-55131 Mainz
Germany

Prof. Dr. H. Böhles
Department of General Paediatrics
Johann Wolfgang Goethe-University
Theodor Stern Kai 7
D-60590 Frankfurt/Main
Germany

Prof. Dr. V. Hesse
Lichtenberg Hospital
Department of Paediatrics
Fanningerstraße 32
D-10365 Berlin
Germany

Dr. C. Kampmann
Children's Hospital
University of Mainz
Langenbeckstr. 1
D-55131 Mainz
Germany

Prof. Dr. E. Kauf
Department of Paediatrics
Friedrich Schiller-University
D-07740 Jena
Germany

Prof. Dr. W. Kienast
Department of Paediatric Cardiology
University of Rostock
Rembrandt-Straße 16/17
D-18055 Rostock
Germany

Prof. Dr. G. Mall
Department of Pathology
Klinikum Darmstadt
Grafenstraße 9
D-64383 Darmstadt
Germany

Priv.-Doz. Dr. T. Marquardt
Department of Paediatrics
University of Münster
Albert-Schweitzer-Str. 33
D-48129 Münster
Germany

Dr. E. Mengel
Children's Hospital
University of Mainz
Langenbeckstr. 1
D-55131 Mainz
Germany

Prof. Dr. Hildegard Przyrembel
Bundesinstitut für Risikobewertung
Thielallee 88-92
D-14195 Berlin
Germany

Priv.-Doz. Dr. R. Santer
University Children's Hospital
Martinistr. 52
D-20246 Hamburg
Germany

Prof. Dr. A. Schmaltz
Department of Paediatric Cardiology
University of Essen
Hufelandstraße 55
D-45147 Essen
Germany

Dr. A. C. Sewell
Department of General Paediatrics
Johann Wolfgang Goethe-University
Theodor Stern Kai 7
D-60590 Frankfurt/Main
Germany

Prof. Dr. W. Sperl
Children's Hospital
Landeskliniken Salzburg
Müllner Hauptstr. 35
A-5020 Salzburg
Austria

Priv.-Doz. Dr. C. F. Wippermann
Department of Paediatrics
University of Mainz
Langenbeckstraße 1
D-55131 Mainz
Germany

Index

A
ACE inhibitors 147
acetate 17
acetyl-CoA 25, 27, 35
acidaemia, organic 22
acid glucosidase 47
– maltase 47, 156
acroparesthesia 124
acylcarnitine 21f., 153f., 157
acyl-CoA 25
– dehydrogenase 27
– –, very-long-chain 27
amino acids 17
amiodarone 147
AMP-activated protein kinase 57
amylo-1,6-glucosidase 49
– deficiency 52
amylo-1,4-1,6-transglucosidase 55
Anderson-Fabry disease 133ff.
angiokeratoma 3, 125, 128, 144
anhidrosis 125
antibodies, antimyolemmal 12
–, antisarcolemmal 12
arterial hypertension 102
ATP 25, 57

B
Barth syndrome 74, 76ff.
biopsy, endomyocardial 3, 7, 8, 11
–, techniques 11

C
carbohydrate-deficient glycoprotein syndrome 85
cardiac energy metabolism 17
– phosphorylase kinase 49
cardiomegaly 19ff., 29, 32, 39f., 44, 58, 60, 154
cardiomyopathy 18, 20, 32, 40
–, adriablastin induced 13
–, anthracycline-induced 6, 11
–, asymmetric hypertrophy 1, 2
–, concentric 145
–, congestive 117
–, differential diagnosis 1, 7
–, dilatative 75
–, dilated form 1, 6f., 13, 18, 21, 31, 72f., 76, 85, 92ff., 96, 104, 110
–, endomyocardial biopsy 3
–, histiocytoid 71f.
–, hypertrophic form 1f., 13, 18, 21, 56f., 72, 74, 76f., 85, 89, 91, 94, 96, 134, 143f., 154
–, idiopathic form 1, 6f.
–, infantile fatal 73
–, infections form 7
–, mitochondrial 155
–, mtDNA mutation 67
–, restrictive form 1
– with genetic disorders 88
carnitine 17, 18, 19, 20ff., 26, 28ff., 43, 78, 153f., 157
–, acyltranslocase 22
– deficiency 4, 13
– palmitoyltransferase 19
–, primary deficiency 18
–, systemic deficiency 18
–, translocase deficiency 21
–, transporter system 17
CDG s. glycosylation, congenital disorders
ceramide storage disease 13
ceramide-trihexoside 3
CK 20
CoA 18
conduction system, changes 104f.
congenital desorders of glycosylation 85, 88, 93f., 96f., 157
coronary arteries 100ff.
– ischaemia 110
cor pulmonale 115
CPT 1 19, 21, 155
CPT 2 19ff., 155
cylcarnitine translocase 25
cystic fibrosis 117ff.
cytochrome C oxidase 72, 77

D
Danon disease 61
debranching enzyme deficiency 155
dehydrogenase
–, long-chain 25
dicarbocylic acids 27, 31
dichloracetate 79
dysostosis multiplex 156

E
endomyocaridal biopsy 78, 93
enoyl-CoA, hydratase 27
enzyme replacement therapy 60, 113, 129f., 147, 150

F
Fabry disease 3f., 13, 123ff., 141, 156f.
fatty acid oxidation 18, 20, 26f.
fatty acids, free 17, 25, 28f., 35, 154
– –, long-chain 19, 35, 153
free fatty acids see fatty acids, free

A
α-galactosidase 127ff.
– A 124, 133, 144, 147
– deficiency 3
Gaucher disease 113ff., 156f.
globotriaosylceramide 124f., 128, 133
β-glucocerebrosidase 113
glucose 17, 25
– utilisation, anaerobic 17
α-glucosidase 52, 60f.
–, deficiency 58ff.

glutathione peroxidase 117
glycogen 12, 47, 57
– debranching enzyme 49
– storage diseases 47, 155
glycogenoses 3
glycosaminoglycans 99, 156, 160
α-glycosidase 51
glycosphingolipids 123, 126, 128, 143
glycosylation, congenital disorders 85, 88, 93f., 96f., 157
GM1 gangliosidosis 156
GM2 gangliosidosis 156

H
heart valve calcification 114
heart valves, alterations 105f.
HELLP 30
– syndrome 89
3-hydroxyacyl-CoA, long-chain 25
3-hydroxybutyrate 154
hypertrophic cardiomyopathy s. cardiomyopathy, hypertrophic
hypertrophy, assymmetrical 135
–, concentric 135
–, eccentric 135
hypoglycaemia 21, 89
–, hypoketotic 18f., 21, 27, 29f., 154
hypohidrosis 125

I
ischaemia, subendocardial 143

K
Kashin-Beck disease 117
Kearns-Sayre syndrome 76, 79
Keshan disease 70, 117
ketone bodies 17
β-ketothiolase deficiency 35

L
lactate 17, 25, 155
lactate/pyruvate ratio 78
lactic acid 78
LAMP1 52
LAMP2 52, 62
L-carnitine 79

lethal infantile mitochondrial disease 76
LHON 76
long-chain fatty acid oxidation 25, 26, 27
long-chain fatty acids see fatty acids, long-chain
long-chain 3-hydroxyacyl-CoA dehydrogenase deficiency 154
lymphocytes, vacuolated 156
lysosomes 47

M
MELAS 75f.
MERRF syndrome 75f.
3-methylglutaconic acid 75, 78
– aciduria 74, 157
methylglutaconnic acid 77
mitochondria 4, 6, 19, 28, 71ff., 78f., 153, 155
mitochondrial cardiomyopathy 7
– disease 67
– myopathy 6
mitochondriopathies 3
mitral valve prolapse 142
mtDNA 70, 73
–, deletion 69, 75
–, mutation 67f., 71, 73
mucopolysaccharidoses 3, 99ff., 156
Mullins-sheath 11
muscle phosphorylase 49
myocardial changes 103f.
myocarditis 7f., 11, 13, 93, 114
myoglobinuria 20, 29f., 32

O
oligosaccharides 156, 160
organic acid, urinary analysis 154
organic acids in urine and plasma 157
oxidative phosphorylating system 67f., 70f., 74f., 77, 79
OXPHOS s. oxidative phosphorylating system

P
pericardial effusion 85, 89
pericarditis 114
phenylketonuria 117ff.
phophorylase kinase 49
phosphofructokinase deficiency 56
phosphomannomutase 91
– 2 89
phosphorylase 47, 49
phosphorylase kinase 56f.
–, deficiency 56f.
phosphorylase-b-kinase deficiency 156
Pompe disease 58, 155
propionic aciduria 157
pyruvate 17, 25

R
respiratory chain 27
– – defect 75, 79
– – enzymes 70f.
rhabdomyolysis 20

S
δ-sarcoglycan 96
selenium deficiency 13, 70, 117ff.
Sengers syndrome 74, 76
serum transferrin 88
sphingoglycolipids 139
sphingolipidosis 113ff.

T
tandem mass spectrometry 21, 153f., 157
2-trans-enoyl-CoA 27
transferrin 85, 92, 94, 157
tyrosinaemia type 1 157

V
vacuolated lymphocytes 61
valproic acid 22
valvular regurgitation 106f., 110, 142
very long-chain acyl-CoA dehydrogenase deficiency 154